# Advances in 21st Century Human Settlements

Indexed by SCOPUS

This Series focuses on the entire spectrum of human settlements—from rural to urban, in different regions of the world, with questions such as: What factors cause and guide the process of change in human settlements from rural to urban in character, from hamlets and villages to towns, cities and megacities? Is this process different across time and space, how and why? Is there a future for rural life? Is it possible or not to have industrial development in rural settlements, and how? Why does 'urban shrinkage' occur? Are the rural areas urbanizing or is that urban areas are undergoing 'ruralisation' (in form of underserviced slums)? What are the challenges faced by 'mega urban regions', and how they can be/are being addressed? What drives economic dynamism in human settlements? Is the urban-based economic growth paradigm the only answer to the quest for sustainable development, or is there an urgent need to balance between economic growth on one hand and ecosystem restoration and conservation on the other—for the future sustainability of human habitats? How and what new technology is helping to achieve sustainable development in human settlements? What sort of changes in the current planning, management and governance of human settlements are needed to face the changing environment including the climate and increasing disaster risks? What is the uniqueness of the new 'socio-cultural spaces' that emerge in human settlements, and how they change over time? As rural settlements become urban, are the new 'urban spaces' resulting in the loss of rural life and 'socio-cultural spaces'? What is leading the preservation of rural 'socio-cultural spaces' within the urbanizing world, and how? What is the emerging nature of the rural-urban interface, and what factors influence it? What are the emerging perspectives that help understand the human-environment-culture complex through the study of human settlements and the related ecosystems, and how do they transform our understanding of cultural landscapes and 'waterscapes' in the 21st Century? What else is and/or likely to be new vis-à-vis human settlements—now and in the future? The Series, therefore, welcomes contributions with fresh cognitive perspectives to understand the new and emerging realities of the 21st Century human settlements. Such perspectives will include a multidisciplinary analysis, constituting of the demographic, spatio-economic, environmental, technological, and planning, management and governance lenses.

If you are interested in submitting a proposal for this series, please contact the Series Editor, or the Publishing Editor:

Bharat Dahiya (bharatdahiya@gmail.com) or
Loyola D'Silva (loyola.dsilva@springer.com)

More information about this series at http://www.springer.com/series/13196

T. M. Vinod Kumar
Editor

# Smart Living for Smart Cities

## Community Study, Ways and Means

Springer

*Editor*
T. M. Vinod Kumar
Kozhikode, Kerala, India

ISSN 2198-2546 ISSN 2198-2554 (electronic)
Advances in 21st Century Human Settlements
ISBN 978-981-15-4602-0 ISBN 978-981-15-4603-7 (eBook)
https://doi.org/10.1007/978-981-15-4603-7



This Springer imprint is published by the registered company Springer Nature Singapore Pte Ltd.
The registered company address is: 152 Beach Road, #21-01/04 Gateway East, Singapore 189721, Singapore

# Contents

# Author Biographies

**T. M. Vinod Kumar** had 49 years of experience in Urban Planning, as a teacher, researcher, and adviser/consultant and worked in India, China, Bhutan, Nepal, Malaysia, Indonesia and Hawaii USA. He was Dean of Studies, Head of the Department of Urban Planning, Head Centre for Systems Studies and Analysis, Centre for GIS and Remote Sensing, and Centre for Urban Studies of School of Planning and Architecture, New Delhi; Visiting Professor National Institute of Technology, Calicut, and Institute of Technology Bandung, Indonesia and Professional Associates, East West Resources Systems Institute Honolulu, Hawaii, Fellow Centre for the Study of Developing Societies, Delhi, Project Manager in Council for Social Development, New Delhi, Regional Program Coordinator at the International Centre for Integrated Mountain Development (ICIMOD) and Planner-Engineer at the Ford Foundation. He is the author of many books and journal articles. He coordinated and edited "Geographic Information System for Smart Cities" (Copal: 2014), "E Governance for Smart Cities" (Springer: 2015) "Smart Economy in Smart Cities" (Springer: 2016), "E-Democracy for Smart Cities" (Springer: 2017) and "Smart Metropolitan Regional Development: Economic and Spatial Design Strategies(Springer-Nature: 2018)", (Springer Nature: 2019) and " Smart Living for Smart Cities Vol 1 and 2" (Springer Nature: 2020).

**Arpita Banerjee** is an urban and regional planner, who has graduated (with Distinction) from the Department of Physical Planning, School of Planning and Architecture, New Delhi. She has been awarded the General Proficiency Award for excellent academic performance and the National Thesis Benefaction Award for Innovative Ideas by the National Organization of Students of Planning (NOSPLAN) in 2018. In 2016, she was awarded the prestigious Apurva Rustagi Scholarship for Meritorious Women in Planning.

She works in the domain of geo-spatial technologies, regional planning and smart city development. She has published papers in International journals and magazines and presented her work in international conferences owing to her thesis on a "Geo-spatial approach to rurban planning. She has worked as a researcher with different organizations for Government of India projects (including the Ministry of Human Resource Development, the Ministry of Tribal Affairs and the Ministry of Housing and Urban Development). She was employed as a Research Associate at the Indian Institute of Public Administration (IIPA) in New Delhi, and is involved in cross-disciplinary research on computerized smart planning for smart regions as an independent researcher. She is an incoming post-graduate student and teaching assistant at the Department of Urban and Regional Planning (DURP), University of Illinois, Urbana Champaign, USA.

**Sweta Byahut** is an Associate Professor in community planning at Auburn University. She earned a PhD in Regional Development Planning from the University of Cincinnati in 2012, a Master's in Planning from CEPT University in 1997, and a Bachelor's in Architecture from M.S. University in India in 1995. Before academia, Dr. Byahut has worked extensively on planning legislation, development regulation, comprehensive planning, regional planning, post-earthquake reconstruction, and inner city revitalization. She continues to research development regulations and urban land management issues in Indian cities. Her interests also include the application of spatial analysis tools in built-form analysis. She has over a dozen peer reviewed publications including in the Journal of Housing and the Built Environment, Environment and Planning B, Journal of Urban Planning and Development, Urban Design International and another forthcoming in the Journal of Urban Design.

**Dr. Amit Chatterjee** An Urban and Regional Planner, Dr. Amit Chatterjee wrote his doctoral dissertation on urban sustainability of Greater Mumbai and its satellite towns. Dr. Amit has a combined experience of more than a decade in teaching, research and industry and presently on the faculty of Planning, School of Planning and Architecture (SPA), Bhopal as Assistant Professor. Before joining academics, Dr. Chatterjee served the industry in various capacities. As Principal Investigator, Dr. Chatterjee was involved in different international collaborative research projects, namely, Co-benefits of Waste Management, Collaborating for Climate, UK-India Capacity Development project for Climate Change and Carbon Management etc. Dr. Amit is currently involved in three of the research and consultancy projects; 'Shelter for All', 'Seven Cluster Development Plans under National Rurban Mission' and Development Plan of Cantonment area. In academics, his research interest is primarily focused on future cities, urban sustainability models and climate change. Besides attending seminars at national and international level, Dr. Chatterjee contributes papers, book chapters to reputed international journals in planning. Dr. Chatterjee edited 11th Issue of International Journal of SPA, Bhopal 'SPANDREL' with the theme titled '*Masking cities Smart and Competitive*'. Dr. Amit is the co-author of forthcoming book titled '*Satellite Towns in Neo-metropolitan Development in India: Lessons from selected cities*' (Springer, 2020).

**Premjeet Das Gupta** is an Architect and Urban Planner teaching as Assistant Professor in School of Planning and Architecture, Bhopal. He has around 12 years of combined professional and academic experience. He has been a part of the DFID-funded KUSP programme in West Bengal from 2005 to 2007. Between 2007 and 2009 he worked in a consulting role with the firms within the ICICI Bank and IL&FS groups. He was an Assistant Planner with Kolkata Metropolitan Development Authority from 2009 to 2013 during which he was involved with transportation and PPP projects. At present he is involved in research on Non Motorized Transport in urban peripheries.

**Sudeshna Ghosh** is an Associate Professor of Regional Planning, in the Department of Geography and Regional Planning at Indiana University of Pennsylvania, USA. Dr. Ghosh earned her PhD in Regional Development Planning from University of Cincinnati, and her Master's degree in City Planning from Indian Institute of Technology, Kharagpur, India. Her research interests and recent publications are focused in the areas of economic development planning, land use modeling, and planning in the developing world. Since joining IUP, she initiated and coordinated study abroad programs to explore urbanization issues in contemporary Indian cities in collaboration with Indian universities.

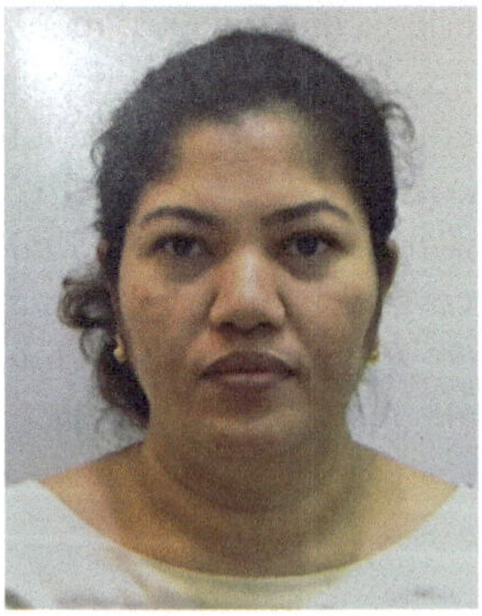

**Ashmita Karmakar** B.Arch, Masters of Town and Regional Planning, PhD Environmental Planning. **Education:** Dr. Ashmita Karmakar graduate in year 1998. She completed her masters in Town and Regional Planning, year 2000 from Indian Institute of Engineering Science and Technology-Shibpur, Kolkata . She further completed her Doctoral studies in Environmental Planning, year 2012 from School of Planning and Architecture, Delhi. **Current Location:** Dubai, UAE. **Area of expertise:** She started her career as an Urban Planner in a consultancy firm in Delhi and then moved to Delhi Development authority as a Planning Officer till year 2007. During this period, she has worked in Several Urban Planning projects of India including Delhi Master Plan and Revitalization of

Old Delhi—Walled City Shahjahanabad. Ashmita has worked for several master planning projects of UAE and other countries as a project manager and coordinator for other infrastructure planning disciplines. In the UAE she has also worked with semi government and Government companies such as Nakheel PJSC, Dubai and Dubai Development Authority respectively. **Project Highlights:** In Nakheel, Ashmita got the opportunity to work as the sole urban planner for Dubai's signature projects such as Palm Jumeirah, The world Island, Deira Island, International City, Jumeirah Park, Jumeirah Village Circle, Jumeirah Island, Jumeirah Island. Her role was to prepare new master plans, alterations to the existing master plans, review and modify parts of master plans, prepare plot affection plans and to prepare development control guidelines for various land uses associated with new and existing plots. She has also worked with the Government entity of UAE—Dubai Development Authority (DDA) and was responsible for reviewing the major master planning projects of UAE and also preparing the master planning and architectural guidelines and policies of the masterplans of Dubai developments under DDA jurisdiction. **Current job profile:** At present she is working with one of the oldest firms of UAE called Khatib and Alami and representing UAE with the role of Proposal manager for the new master planning and land scape architecture projects and coordinator for infrastructure projects for Urban Planning and Landscape architecture Projects in Middle east and Asia. She is also working on the upcoming projects of UAE. **Publications:** She has published papers, articles and chapters on Urban planning, Environmental Planning and Geographical Information System. Her latest Chapters are contributed to Smart City Dubai and are published by Springers Publishers, Singapore.

**Ar. Priyanka Kumar** is Urban Planner in the Regional Centre for Urban and Environmental Studies (RCUES) Lucknow, under the Ministry of Housing and Urban Affairs, Government of India. Currently, she is working for central government schemes such as Smart City Mission, Swachh Bharat Mission, AMRUT, and Pradhan Mantri Awas Yojna. She obtained her Bachelor of Architecture from Kurukshetra University (with first position), in 2016 and earned Best Thesis Award (Thesis titled "Development in Air Space") in Faculty of Architecture, RPIIT Campus Karnal affiliated from Kurukshetra University, Kurukshetra. She also obtained Master in Planning with specialization in Urban Planning (first class with distinction) from the School of Planning and Architecture (SPA), New Delhi in 2018. She is pursuing her PhD in Department of Transport Planning at School of Planning and Architecture (SPA), New Delhi. She "earned" PG Diploma in Public Administration (PGDPA) from Lucknow University, Lucknow in 2019. She has won Young Achievers Award in 2015 from National Integrated forum of Artists and Activists. She has 2 years of experience related to Urban Planning and Public Administration and involved in various projects of Urban-Regional Planning, Infrastructure Planning, Monitoring and Evaluation, Reforms and Urban Governance. She has conducted numerous capacity building programmes and research studies related to various urban schemes and policies at RCUES Lucknow. She has also worked as a Research Associate at the Indian Institute of Public Administration (IIPA) in New Delhi. She has worked as Intern at Ministry of External Affairs, New Delhi. She is also on Editorial Advisory board for various journals. She has several articles/papers published in national and international journals, workshops and seminars. She has been invited as Keynote Speaker/Guest Faculty/External Jury Member in many conferences, Training programs, institutions and colleges respectively. She has been an active member of different professional organizations and various NGOs. She is enrolled for KATAJA Theories and Research in Logistics and Supply Chain Management summer course at Supply Chain Management & Social Responsibility, Hanken School of Economics, Helsinki, Finland.

**Dr. Kusum Lata** is Associate Professor (Urban and Regional Planning) in the Centre for Urban Studies at Indian Institute of Public Administration (IIPA), New Delhi under Department of Personnel and Training, Government of India. She successfully conducted the 44th Advance Professional Programme in Public Administration, the flagship program of IIPA as Program Co-Director. She is Engineer Planner with Bachelor Degree in Civil Engineering from National Institute of Technology (NIT) Raipur, Chhattisgarh (formerly Govt. Engineering College, Raipur, M.P.) in 1987 and Master of Planning from the School of Planning and Architecture, (SPA) New Delhi in 1991. She earned her Ph.D. from SPA, New Delhi in 2004 with established research on Monitoring Urban Development, case study: water supply system of Ghaziabad. She has over 25 years of experience related to planning and database management and involved in various projects in Urban Planning, Infrastructure Planning and Management, Monitoring and Evaluation, Application of GIS for Management of Utilities and Services; and Application of MIS to Governance. She has also worked in government, non-government organizations and consultancy firms on various projects and in different capacities ranging from project associate to project manager and consultant. For more than one decade at IIPA, she has been conducting numerous capacity building programmes and research studies related to urban sector. She has couple of books to her credit and several articles published in national and international workshops and seminars.

**Calvin Masilela** is a tenured full Professor in the Department of Geography and Regional Planning at Indiana Univerity of Pennsylvania, USA. He holds a Ph.D. in Environmental Design and Planning from Virginia Tech, USA. Before joining the department in 2001, Professor Masilela taught at West Virginia University for eight years where he was granted tenure and promotion to Associate Professor. He currently serves as Director of the Bachelor of Science in Regional Planning Program as well as Director of the Upward Bound Math and Science Program. His research interests range across a variety of issues that span African development; Land use planning; Rural development and health; Urban agriculture; and

Urban management. Professor Masilela's scholarly works on planning education, land reform, land use, urban agriculture, development and urbanization issues has appeared in the African Geographical Review, Habitat International, Journal of Geography, Planning and Environment B, Regional Development Dialogue, Small Town, Third World Planning Review, The Pennsylvania Geographer, and The Wiley Blackwell Encyclopedia of Urban and Regional Studies. He is also a regular proposal and grant reviewer for Fulbright-Hays, and National Science Foundation as well as manuscript reviewer for the African Geographical Review, Journal of Urbanism, Journal of Planning Education and Research. In the past, he has reviewed manuscripts for the Journal of Economic Geography, Social and Medicine, Hong Kong Papers in Design and Development, Canadian Journal of African Studies, and Journal of Urban Studies.

**Mr. N. K. Paulose** s an urban geographer, he holds a post-graduate degree in Geography from the University of Kerala and earned another Post graduation in Urban and Regional Planning from CEPT University, Ahmadabad. He was awarded coveted commonwealth Scholarship by Commonwealth Scholarship Commission to study Masters in Transport Planning from the University of Leeds, the UK in 2014. He joined School of Planning and Architecture (SPA), Bhopal in 2011 as an assistant professor. He has close to ten years of consultancy and teaching experience in the field of Urban Planning. His primary area of expertise includes transport policy, public transport and low carbon mobility, climate change and urban management. Currently, he is pursuing his Doctoral Research at SPA Bhopal.

**Ashikha Raoof** B.Tech in Civil engineering, M.Plan in Urban Planning. **Education:** Ashikha Raoof graduated in the year 2014. She completed her master's in urban planning, in the year 2017 from National Institute of Technology, Calicut. **Current Location:** Dubai, UAE. **Area of expertise:** She has started her career as a development consultant in a consultancy firm in Dubai, UAE. In her role, she has worked on various projects in residential development, master planning, infrastructure development and sustainability. Her area of expertise is project development and management. **Project Highlights:** In Middle East Investment and Management Consultancy, Ashikha got the opportunity to be part of various signature projects such as Fujairah Master Plan 2040, Ajman Air Quality strategy, financial feasibility studies for residential projects and other infrastructure projects in solid waste management and wastewater treatment. **Current job profile:** At present, she is working with Middle East Investment and Management Consultancy, a boutique consultancy firm in Dubai, UAE. As a Development Consultant, she is engaged in projects from early stages of inception and is responsible for the managing the design and execution stages of the project. **Publications:** She has published papers on Importance of comprehensive planning in urban planning and the phenomenon of urban heat islands.

**Gaurav Vaidya** is an Infrastructure Planner and Asst. Professor in School of Planning and Architecture, Bhopal. He has around 11 years of professional experience in industry and academics as well. He had worked for more than 4 years (during 2009–2013) as Urban Planner in the Urban Development and Urban Housing Department, Govt. of Gujarat on various assignments, some of them are; preparation of Gandhinagar Development Plan, Ahmedabad Gandhinagar Comprehensive Mobility Plan, Clean Green and Solar Gandhinagar Master Plan, 3 Town Planning Schemes, 12 City Development Plans, Detail Project Reports for Municipal Water Supply and Urban Poor Housing Projects, Urban Reforms Appraisal and Capacity Building Programs for Municipalities of Gujarat etc. As civil engineer he executed various small scale civil engineering projects during 2004–2007 with the capacity of Asst. Consultant Engineer. At present he is involved in the research of "Management of urban sanitation services", "Regional-urban infrastructure linkages" and "Green infrastructure development".

# Introduction

# Smart Living for Smart Cities

**T. M. Vinod Kumar**

**Abstract** Smart living is one of six components of a smart city. Other five are smart people, smart mobility, smart economy, smart environment, and smart government. All smart living activities have also these six components integrated with smart city system. Cities have their own characteristics based on the size whether it is a metropolis, megacity, meta city or small or very big continental settlement. This is based on its unique cultural system and ecological system. All aspects of living in a city can be traced to one major aspect which considerably influences daily life. It can be religion, for example, Islam in Saudi Arabia, Buddhism in Tibet, Judaism in Israel or Christianity in Vatican or it can be a dominant aspect of life which the cultural system has thrived to develop in the past such as for example health as indicated by high performance of related time series statistics and so on. Then the concept of smart is presented illustrating with the smart home which differs greatly with the smart aspect of the city. The design of smart living for smart cities can be centred around this dominant aspect which I call domain and some of the essential features of such domain is briefly presented in this chapter. The chapter then focuses on design for smart living in for smart cities based on a domain approach, IOT and ICT system design for smart living, and design of E-Democracy and E-Governance system for a smart living for smart cities. This will provide for a total design for smart living in a smart city. This is briefly presented in this chapter.

**Keywords** Smart city · Smart living · Smart living domains · Examples · Smart home and smart cities differences · Smart living design · Design of ICT and IOT system for smart living · The design of E-Democracy system for smart living · Design of E-Governance system for smart living · A smart community for a smart living

T. M. Vinod Kumar (✉)
School of Planning and Architecture, New Delhi, India
e-mail: tmvinod@gmail.com



T. M. Vinod Kumar (ed.), *Smart Living for Smart Cities*,
Advances in 21st Century Human Settlements,
https://doi.org/10.1007/978-981-15-4603-7_1

## 1 The Concept of the Smart City

The smart city is conceived around the concept of six components-systems smart people, smart mobility, smart economy, smart environment, smart government and smart living working together in an integrated manner as discussed in the following para 1.2. This way of working around these concepts transforms any city into a smart city. The level of the smart city is the level in which the six components have progressed in its working in these cities. There is no end or saturation in any of these six components if innovative practices are emerging from creative people in future. Every city has some level of achievements in these components in a very partial and non-integrated manner.

The concept of smart is a product of the twenty-first century. We have smartphone which converts our single function dumb phone to a multifunctional capability of a powerful but tiny computer in your pocket that perform all tasks of communication in a more efficient and creative way using all modes of communication and provides you with many extras like video messaging, SMS/text messaging, email messaging, music player in the background, camera, word processing, spreadsheets and so on. The smart plug or smart switch in a smart home is no more the dumb plug doing the assigned function but can be activated by Amazon Echo talk commands to switch on the TV attached to it or switch off. The smart plug monitors the heat generated in the plug and can be programmed to cut off the power when the heat generated to a certain unacceptable level. It can be programmed to switch on a certain time and switch off a certain time. You can then add by design many more functions to smart plug/switch and only your imagination limits the functional addition. In smart devices, you can make many of them work together in a group triggered by motion sensors, thermal and humidity sensors and light sensor and can create scenes you want at certain times predetermined where some or many devices work together as per needs. The smart added as a prefix to the six components are like what has been discussed above but in a more complex way useful for living.

Rightfully, authors of the book "Geographic Information System for Smart Cities" [1] defined smart cities as follows. "Smart city is a knowledge-based city that develops extraordinary capabilities to be self-aware, how it functions 24 h and 7 days a week and communicate, selectively, in real time knowledge to citizen end users for a satisfactory way of life with easy public delivery of services, comfortable mobility, conserve energy, environment and other natural resources, and create energetic face to face communities and a vibrant urban economy even at a time there are National economic downturns [1].

The smart city is not an end state of a city but constantly evolving as the creative and innovative imagination of the citizen evolve with the more innovative functionality of the city added to its six components as per the needs.

The six components system of the smart cities were discussed in many books of the smart city series which was edited by the author [1–6]. These components are discussed here to place the smart living component in the right perspective.

## 2 The Six Smart City Components [3]

The six components of a smart city are diagrammatically presented below. The author gives more prominence to people in this diagram since people assume a proactive role and drives the creation and functioning of smart cities and assumes considerable importance in the smart city concept. The lower of two hierarchy where people appear is at the household level and community level. In the community level involvement of people in smart city converts themselves to the smart community the basic building block of the smart city (Fig. 1).

### *2.1 Smart People*

'Smart People', the fundamental building block of a Smart City System, require many crucial attributes as given.

(1) Smart people excel at what they do professionally.
(2) Smart people have a high Human Development Index [1].
(3) A smart city integrates its universities and colleges into all aspects of city life.
(4) It attracts high human capital, for example, knowledge workers.

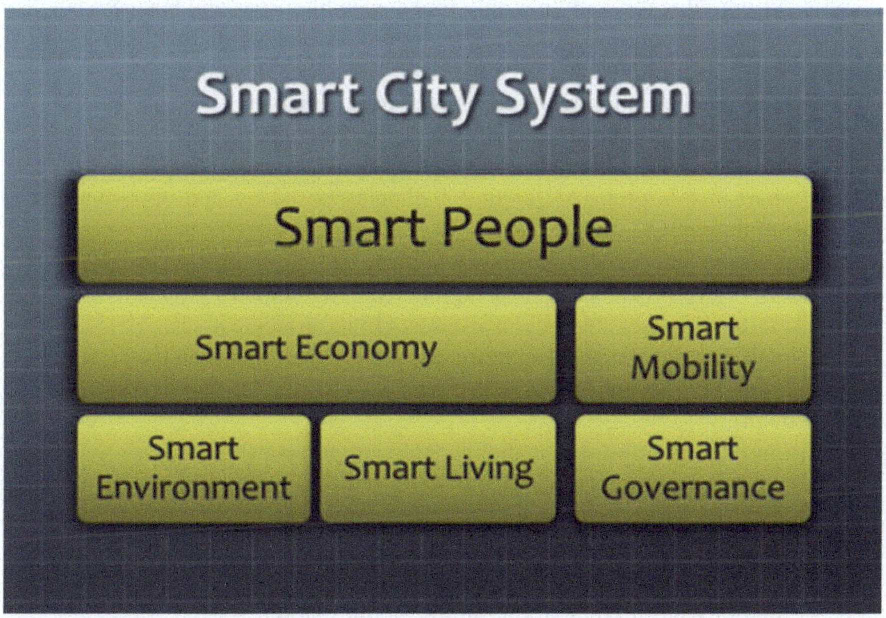


**Fig. 1** Smart city system building blocks. *Source* Vinod Kumar, p. 19 [3]

(5) A smart city maintains high Graduate Enrolment Ratio and has people with a high level of qualifications and expertise.
(6) Its inhabitants opt for lifelong learning and use e-learning models.
(7) People in a smart city are highly flexible and resilient to the changing circumstances.
(8) Smart city inhabitants excel in creativity and find unique solutions to challenging issues.
(9) Smart people are cosmopolitan, are open-minded, and hold a multicultural perspective.
(10) Smart people maintain a healthy lifestyle.
(11) Smart people are actively involved in their city's sustainable development, its efficient and smooth functioning, its upkeep and management, and making it more liveable.

## *2.2 Smart City Economy*

'Smart City Economy', the second building block, requires the following attributes.

(1) A smart city understands its economic DNA.
(2) A smart city is driven by innovation and supported by universities that focus on cutting-edge research, not only for science, industry, and business but also for cultural heritage, architecture, planning, development, and the like.
(3) A smart city highly values creativity and welcomes new ideas.
(4) A smart city has enlightened entrepreneurial leadership.
(5) A smart city offers its citizens diverse economic opportunities.
(6) A smart city knows that all economics works at the local level.
(7) A smart city is prepared for the challenges posed by and opportunities of economic globalization.
(8) A smart city experiment supports and promotes the sharing economy.
(9) A smart city thinks locally, acts regionally, and competes globally.
(10) A smart city makes strategic investments in its strategic assets.
(11) A smart city develops and supports compelling national brand/s.
(12) A smart city insists on balanced and sustainable economic development (growth).
(13) A smart city is a destination that people want to visit (tourism).
(14) A smart city is nationally competitive on selected and significant factors.
(15) A smart city is resourceful, making the most of its assets while finding solutions to problems.
(16) A smart city excels in productivity.
(17) A smart city has high flexibility in the labour market.
(18) A smart city welcomes human resources that enhance its wealth.
(19) A smart city's inhabitants strive for sustainable natural resource management and understand that without this its economy will not function indefinitely.

## 2.3 *Smart Mobility*

'Smart Mobility', the third building block of a Smart City System, includes the following features.

(1) A smart city focuses on the mobility of people and not only that of vehicles.
(2) A smart city will advocate walkability and cycling.
(3) A smart city has vibrant streets (at no additional cost).
(4) A smart city effectively manages vehicular and pedestrian traffic, and traffic congestion.
(5) A smart city has pleasurable (bicycle) routes.
(6) A smart city has balanced transportation options.
(7) A smart city will have a mass rapid transit system, such as metro rail, light metro, monorail, or 'sky train' for high-speed mobility.
(8) A smart city will have integrated high-mobility system linking residential areas, workplaces, recreational areas, and transport notes (e.g. bus/railway station/s and airport).
(9) A smart city will practice high-density living, such that the benefit of high-speed mobility is uniformly available.
(10) A smart city has seamless mobility for differently abled (often incorrectly called, disabled) people.

## 2.4 *Smart Environment*

'Smart Environment', the fourth building block, has the following attributes.

(1) A smart city lives with and protects nature.
(2) A smart city is attractive and has a strong sense of place that is rooted in its natural setting.
(3) A smart city values its natural heritage, unique natural resources, biodiversity, and environment.
(4) A smart city conserves and preserves the ecological system in the city region.
(5) A smart city embraces and sustains biodiversity in the city region.
(6) A smart city efficiently and effectively manages its natural resource base.
(7) A smart city has recreational opportunities for people of all ages.
(8) A smart city is a green and blue city.
(9) A smart city is a clean city.
(10) A smart city has adequate and accessible public green spaces.
(11) A smart city has an outdoor living room. Unlike the indoor living room in houses where we meet others, outdoor living rooms are aesthetically designed intimate, active, and dynamic urban realms where people meet face to face for a culturally and recreationally rich and enjoyable contact as part of living and work.

(12) A smart city has distinctive and vibrant neighbourhoods that encourage neighbourliness and a spirit of community.
(13) A smart city values and capitalizes on scenic resources without harming the ecological system, natural resources, and biodiversity.
(14) A smart city has an integrated system to manage its water resources, water supply system, wastewater, natural drainage, floods and inundation, especially in the watersheds where it is located, especially in view of the (impending) climate change.
(15) A smart city focuses on water conservation and minimizes the unnecessary consumption of water for residential, institutional, commercial, and industrial use, especially in the arid and semi-arid areas.
(16) A smart city has an efficient management system for the treatment and disposal of wastewater, and reuse of treated wastewater, particularly in the arid and semi-arid areas.
(17) A smart city has an efficient management system for the collection, treatment, and disposal of industrial wastewater.
(18) A smart city has an integrated and efficient management system for the collection, transfer, transportation, treatment, recycling, reuse, and disposal of a municipal, hospital, industrial, and hazardous solid waste.
(19) A smart city has an efficient system to control air pollution and maintain clear air, especially in the air sheds where it is located.
(20) A smart city has an efficient and effective system for disaster risk reduction, response, recovery, and management.
(21) A smart city has and continually upgrades its urban resilience to the impacts of climate change.
(22) A smart city can create a low-carbon environment with a focus on energy efficiency, renewable energy, and the like.

## 2.5 *Smart Living*

'Smart living', the fifth building block of a Smart City System, includes the following features.

(1) A smart city has strong and shared values.
(2) A smart city records and celebrates local history, culture, and nature.
(3) A smart city has a vibrant downtown, 24 h and 7 days a week.
(4) A smart city can provide the necessary safety and security to women, children, and senior citizens.
(5) A smart city improves the urban way of life.
(6) A smart city builds natural and cultural assets to build a good quality of life.
(7) A smart city not only understands the big picture of urban liveability but also pays attention to small details.
(8) A smart city has high-quality open and accessible public spaces.

(9) A smart city has high-quality public services and amenities.
(10) A smart city is an ideal place of living, especially for women, children, and senior citizens.
(11) A smart city organizes festivals that celebrate people, life, and nature in the city.
(12) A smart city has a ritual event (or more) that symbolizes the values and aspirations of the community.
(13) A smart city celebrates and promotes art, cultural, and natural heritage in the city.
(14) A smart city engages artists to improve and enrich the aesthetics of daily life of the city.

## *2.6 Smart Governance [2]*

'Smart Governance', the six and final building block, has the following attributes.

(1) A smart city practices accountability, responsiveness, and transparency (ART) in its governance.
(2) A smart city uses big data, spatial decision support systems and related geospatial technologies in urban and city regional governance.
(3) A smart city constantly innovates e-governance for the benefit of all its residents.
(4) A smart city constantly improves its ability to deliver public services efficiently and effectively.
(5) A smart city practices participatory policymaking, planning, budgeting, implementation, and monitoring.
(6) A smart city has a clear sustainable urban development strategy and perspectives known to all.
(7) A smart city utilizes creative urban and regional planning with a focus on the integration of economic, social, and environmental dimensions of urban development.
(8) A smart city features an effective, efficient, and people-friendly urban management.
(9) A smart city practices E-Democracy to achieve better development outcomes for all.
(10) A smart city embraces a Triple Helix Model in which government, Academia and Business/Industry practice changing roles in Governance.

## 3 From Living in Cities to Smart Living in Smart Cities

A City provides with a complex space for all living. Wellbeing in a city for an individual is often satisfactory, sometimes unsatisfactory and often threatened with city environment that is largely unfamiliar. No individual alone can control all the external environment of the city he is part of.

Most of the inhabitant has migrated to the city from other cities or villages far and near this generation or many generations back. The house they live is different from what he had earlier, the city environment is again different, but they all carry the baggage of their own familiar culture (regularity of behaviour) of living practised by their elder generations with them which forms the basis of their living in a city. With what city provides for he balances his culture of living with familiar along with the unfamiliar.

All living search for a place or neighbourhood where similar population with identical culture are living in large numbers that provide for special services and facilities such as churches and temples where he can meet people with similar culture, shopping areas where familiar food is marketed, cultural centres where art forms he is familiar with is practised and so on which satisfies his traditional culture for similar people if there is a choice and this balancing act with familiar to unfamiliar culture reaches a higher level of adjustment such that familiar culture is never forgotten. The community he was surrounded in his past habitat who supported him in his difficulty is no more there. There is a need to form new associations of the community around him to replicate what he had earlier. There are formal as well as informal institutions locally and he would like to make it smart and innovative.

A city becomes a Smart city when it reaches a certain level of attainment in the six components discussed in Sect. 2. Some people simplify this accomplishment or evolution of a city as ICT enriched city in all walks of life. Some may add IOTs along with ICT and even say the city is smart based on the number of IOT used per sq.km. of the urban area. What ICT and IOT provide for is additional awareness of city for the conduct of living and that living can be called smart living. Just like Smart city, Smart living does not happen with the addition of per capita income like the happiness of an individual do not add with the addition of money he earns alone in a city in comparison to the village where he came from. So smart city is not the addition of ICT and IOT per sq. km. of urban land.

Just addition of ICT and IOT is not enough for a smart city to be fully functional; there should be a community that practices E-Democracy to meet their living needs and E-Governance for the perfect implementation of Central and State Government legislation as related to living.

A deliberate design of smart living is required to be executed by the community. How to design such smart living is a challenge this book address.

### *3.1 The Smart Awareness for Smart Living in Smart Cities*

The smart city first makes it self-aware making use of a variety of means which will be discussed in para 1.4 and convert city into a unique knowledge-based settlement constantly being updated about itself 24 h and 7 days a week to conduct smart living. In doing so the unique essence of the city is highlighted which cannot be duplicated in another city and need to be celebrated by all. This unique characteristic of the city that changes from every settlement is the unique ecological and cultural city system of the settlement [1].

The smart city is, therefore, a unique but complex by a mixture of cultural groups and varying spatially ecological and cultural system. Because of the unique ecology of Smart Cities, a generalised model of the city is not possible. So also, a generalised smart solution for a smart city is not workable using ICT and IOT. The spatial and ecological structure of built space forming land use and transportation system also varies and as other ecological elements of the city. However, these ecological elements are ideal candidates for dealing with the smart city conceptualisation and for cyberspace interventions in real time using the Internet of Things and ICT.

The smart city is also having unique cultural characteristics. There are innumerable factors that influence the culture of a people in a smart city. The culture of the city is determined by the regularity of behaviour of inhabitants of smart cities. These factors determine the cultural attributes of smart cities. These cultural factors also have a bearing on smart city functioning and will influence the functioning and managing of smart cities in 24 h and 7 days a week functioning. Also, there is a cultural element that influences the deployment of smart technologies. While water supply can be efficiently provided for by smart technology, water use pattern in a smart city which aims for water conservation is culturally conditioned and require cultural adaptation to be water conserving. In any country where there is open air defecation is culturally prevalent then the use pattern of the built toilet must be by cultural transformation if it all it is to be used.

## 4 Cultural System and Basic Ekistics Principles by Constantinos Doxiadis [7–13]

Constantinos Doxiadis [7] first suggested that human settlements are susceptible to systematic investigation. He said, to create the cities of the future including smart cities of tomorrow, he said we need to systematically develop a science of human settlements. This science he called Ekistics, which will take into consideration the principles, a person considers when building his settlements, as well as the evolution of human settlements from dumb city to the smart city today through history in terms of size and quality. The target is to build a smartcity, which respects human dimensions for smart living. Since there is no point in resisting development, we should try to accommodate technological evolution and the needs of man within the same settlement using ICT and IOT.

Study of the city for me is the study of this unique ecological and cultural system and smart living is the celebration of the unique ecological and cultural system that constitutes the city by people. As the size of settlement increases so also the complexity. Then, a better way of study is the classification of human settlements into parts and then study it. Looking at the global cities Doxiadis divided the human settlement according to the population on a logarithmic scale of 15 parts/population size. These 15 levels can be used for the study of the ecological and cultural system of the settlement. It is tabulated in Table 1. To achieve a proper classification, by sizes, of all human settlements, we should start he said with the smallest units. The smallest one is the man himself as an individual. This spatial unit includes the individual, his clothing, and certain furniture, like his chair. It can include wearable IOTs. The second unit is also very well defined; it is the space which belongs to him alone or is shared under certain circumstances with a few others; that is, his personal room. The third unit, the family home, is well defined also if we have families. The fourth unit is a group of homes which corresponds to the patriarchal home of earlier days and probably to the unit of the extended family of our day; this is the unit that children need most, mothers need mainly because of the children, and fathers need

**Table 1** Logarithmic Ekistics scaling of settlement by Doxiadis

| S. No. | Unit | Population (Numbers) |
|---|---|---|
| 1 | Anthropos/Man | 1 |
| 2 | Room | 2 |
| 3 | House | 5 |
| 4 | House Group/Hamlet | 40 |
| 5 | Small Neighbourhood/Village | 250 |
| 6 | Neighbourhood | 1500 |
| 7 | Small Town/Polis | 10,000 |
| 8 | Polis/City | 75,000 |
| 9 | Small Metropolis | 500,000 |
| 10 | Metropolis | 4 million |
| 11 | Small Megalopolis | 25 million |
| 12 | Megalopolis | 150 million |
| 13 | Small Eperopolis | 750 million |
| 14 | Eperopolis | 7500 million |
| 15 | Ecumenopolis | 50,000 million |

*Note* UN-HABITAT statistics define Metropolis with 1 million population urban agglomeration, Megacities with 10 million and Meta cities with 20 million population and above. Census of India classifies Urban Agglomeration above 1 million as Metropolitan area if perhaps not directly for themselves, because they are interested in the satisfaction and happiness of both mother and children. Doxiadis has defined four units; of these, the first three are very clearly defined, physically and socially, and the fourth can be conceived of as a social unit.

Beyond this point he said, we do not have a clear-cut definition of any unit until we reach the largest one possible on this earth—that is, the systems of human settlements of the whole planet. Thus, we have five basic units, four at one extreme of our scale and one at the other. No other well-defined unit exists today, except for statistically defined units which are arbitrary, as may be seen from the differences in the official definitions from country to country. If we turn back in history he said, we find, however, that, throughout the long evolution of human settlements, people in all parts of the world tended to build an urban settlement which reached an optimum size of 50,000 people and physical dimensions such that everyone was within a 10-min distance from the centre [7]. There is no question that, for people who depend on walking as a means of locomotion, this unit is the optimum one from the point of view of the movement and social interaction through direct contacts between people. Also, experience has shown that, for people who can walk, it is a maximum one from the standpoint of aesthetics; on foot. It is also perhaps an optimum one from the social point of view; for example, he said, Pericles in ancient Athens could get a reasonable sample of public opinion by meeting 100–150 people while walking from his home to the Assembly. Thus, we now have four units at the beginning of the scale, one larger, one somewhere beyond them, and one at the end—a total of six. How can we complete the scale? This can be achieved, for example, if we think of units of space measured by their surface and increase their size by multiplying them by 7. Such a coefficient is based on the theory, presented by Christaller [15] he said, that we can divide space in a rational way by hexagons—that one hexagon can become the centre of seven equal ones. Similar conclusions can be reached if we think of organization of population, movement, transportation, and so on. Thus, the basic units are defined as units No. 1 (man), No. 2 (room), No. 3 (home), No. 4 (a group of homes), No. 8 (traditional town), and No. 15 (universal city), and a systematic subdivision defines the others. All these units can also be classified in terms of communities, of kinetic fields (for pedestrians, from a motor vehicle, and so on).

Doxiadis Ecumenopolis is as per his sketch below (Fig. 2).

Doxiadis, further stated that We cannot acquire proper knowledge about our villages, towns and cities unless we manage to see the whole range of the man-made systems within which we live, from the most primitive to the most developed ones—that is, the whole range of human settlements. Our subject, the whole range of human settlements, is a very complex system of five elements—nature, man, society, shells (that is, buildings), and networks. It is a system of natural, social, and man-made elements which can be seen in many ways—economic, social, political, technological, and cultural. For this reason, only, the widest possible view can help us to understand it. He postulated the principles of human settlement.

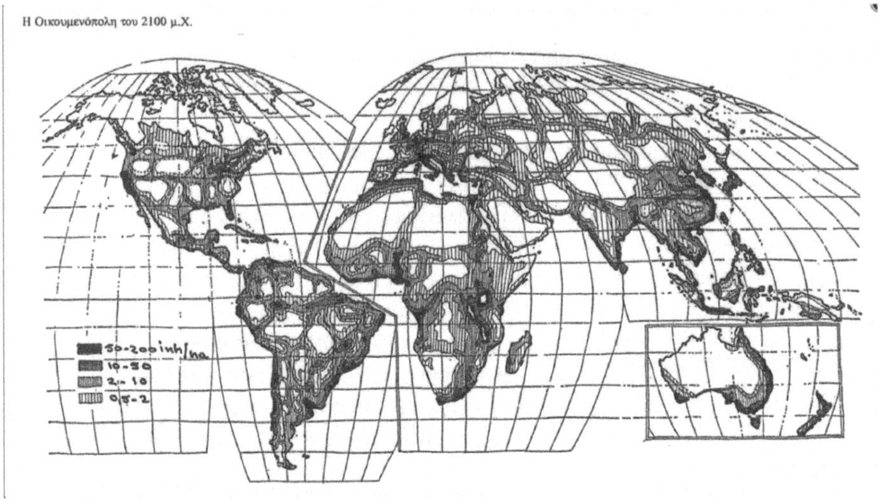


**Fig. 2** Ecumenopolis. *Source* Doxiadis

### *4.1 The Five Ekistics Principles of Doxiadis [7]*

In shaping his settlements, man has always acted in obedience to five principles. As far as Doxiadis know, this has always been true, and he has not found any cases which prove the opposite.

The first principle is the maximization of man's potential contacts with the elements of nature (such as water and trees), with other people, and with the works of man (such as buildings and roads). This, after all, amounts to an operational definition of personal human freedom. It is because of this principle that man considers himself imprisoned, even if given the best type of environment if he is surrounded by a wall without doors. In this, man differs from animals; we do not know of any species of animals that try to increase their potential contacts with the environment once they have reached the optimum number of contacts. Man, alone always seeks to increase his contacts. ICT enables it in a big way.

The second principle is the minimization of the effort required for the achievement of man's actual and potential contacts. He always gives his structures the shape or selects the route, that requires the minimum effort. The shortest path algorithm is built into his mind.

The third principle is the optimization of man's protective space, which means the selection of such a distance from other persons, animals, or objects that he can keep his contacts with them (first principle) without any kind of sensory or psychological discomfort. This must be true at every moment and in every locality, whether it is temporary or permanent and whether a man is alone or part of a group. This has been demonstrated very well, lately, for the single individual, by anthropologists such as Hall [14] and psychiatrists such as Kinzel [15], and by the clothes man designs for himself, and it may be explained not only as a psychological but also asa physiological problem if we think of the layers of air that surround us [12] or the energy that we represent. The walls of houses or fortification walls around cities are other expressions of this third principle.

The fourth principle is the optimization of the quality of man's relationship with his environment, which consists of nature, society, shells (buildings and houses of all sorts), and networks (ranging from roads to telecommunications). This is the principle that leads to order, physiological and aesthetic, and that influences architecture and, in many respects, art.

Finally, and this is the fifth principle, man organizes his settlements to achieve an optimum synthesis of the other four principles, and this optimization is dependent on time and space, on actual conditions, and on man's ability to create a synthesis. When he has achieved this by creating a system of floors, walls, roofs, doors, and windows which allows him to maximize his potential contacts (first principle) while minimizing the energy expended (second principle) and at the same time makes possible his separation from others (third principle)and the desirable relationship with his environment (fourth principle), we speak of "successful human settlements". What we mean are settlements that have achieved a balance between man and his man-made environment, by complying with all five principles. All five principles constitute the basic principles of the cultural system in a city.

Doxiadis concentrated mostly on the cultural system of human settlements while Patrick Geddes expanded it to ecological which he calls "nature and culture". Doxiadis main contribution as far as the objectives of this book is to show there are several levels of complexity in human habitat, some 15 levels and merit separate approaches based on increasing cultural complexity. This means smart home ICT and IOT technologies cannot be extrapolated to the metropolis, mega city and meta city in the same way. Another important question is can his limit of 50,000 population city be expanded to a meta city with 20 million plus population with the help of artificial intelligence, IOT system and ICT system well designed to suit the need he mentioned in his 5 principles? Finally, it can be concluded that his 5 principles have the potential to be used in the concept of smart living in smart cities.

## 5 Ecological Planning and Design and Bioregionalism by Patrick Geddes [16–20]

Geddes advocated the interdependence of culture and nature and merits in our approach to smart cities and smart living. He advocated the design approach in which one materially adopt people and livelihood to specific challenges and opportunities of a place they inhabit. This resulted according to him a marital expression of culture in spatial design in nature.

Patrick Geddes is correctly described as one of the founders of the modern town and regional planning. His theoretical ideas have influenced much subsequent planning practice, regional economic development and environmental management. His focus on the triad "Place–Work–Folk" is fundamental to contemporary debates and research into regionalism and locality, economic and community regeneration, environmental quality and sustainable development and social inclusion.

Geddes' work was based on his fundamental principle of 'Place, Work and People' as he believed that geography; economics and anthropology were related, yielding a single chord of social life. He saw sociology as a quintessentially interdisciplinary subject that was essentially the science of man's interaction with a natural environment, and improved urban planning was one of the key applications of sociology. Geddes revived the Old Town of Edinburgh using his ground-breaking concepts of 'diagnostic surveys' and 'conservative surgery', which he went on to implement in town planning projects across Scotland, India and the Middle East. In fact, Tel Aviv is a city whose core was entirely built around Geddes' plan. He also introduced the concept of 'region' to architecture and planning and coined the term 'conurbation', the present metropolis.

Many have been influenced by his work, including urban theorists such as Lewis Mumford. Geddes also set the foundation for future urban thinkers such as Jane Jacobs and the New Urbanism movement which focuses on walkable neighbourhoods, a variety of housing and occupation types.

Patrick Geddes explained an organism's relationship to its environment as follows:

"The environment acts, through function, upon the organism and conversely the organism acts, through function, upon the environment", he says in his book Cities in Evolution, published in 1915 [16, 17]. In human terms, this can be understood as a place acting through climatic and geographic processes upon people and thus shape them. At the same time, people act, through economic processes such as farming and construction, in a place and thus shape it. Thus, both place and folk are linked and through work are in constant transition.

Geddes first published his idea of the valley section in 1909 to illustrate his idea of the 'region city'. The region is expressed in the city and the city spreads the influence of the highest level in the region. To put it another way, Geddes said that "it takes a whole region to make the city". The valley section illustrated the application of Geddes's trilogy of 'folk/work/place' to the analysis of the region. The valley section is a complex model, which combines physical condition—geology and geomorphology and their biological associations—with so-called natural or basic occupations such as miner, hunter, shepherd or fisher, and with the human settlements that arise from them.

The constellation theory was also coined by Sir Patrick Geddes, which are not economically, politically, socially equal come together in developing a whole region. This theory is mostly used for an administrative purpose in all countries worldwide. Such a theory is most prominently used because planning cities in a shape pattern are not possible in Today's times.

The term “conurbation” was coined in 1915 by Patrick Geddes in his book Cities in Evolution [16, 17]. Internationally, the term “urban agglomeration” is often used to convey a similar meaning to “conurbation”. The conurbation is—waves of population inflow to large cities, followed by overcrowding and slum formation, and then the wave of backflow—the whole process resulting in amorphous sprawl, waste, and unnecessary obsolescence. A conurbation is a region comprising several cities, large towns, and other urban areas that, through population growth and physical expansion, have merged to form one continuous urban and industrially developed area. UN-HABITAT in 2016 renamed it as City avoiding the tag region.

Geddes concept of life is one that connects culture with nature. He believes that nature conservation in cities, technology acts as a vehicle to meet human needs in a city that integrate ecological and social processes to create a healthy environment. The current competitive outlook will give way to cooperative outlook at the regional, international and global scale. He was foreseeing regional economy based on renewable energy.

Geddes finally gave a diagrammatic representation of living. This is based on Facts, Thoughts, Acts and Deeds. All these four aspects are connected one with the other to form life. A closer study of this diagram shows he has incorporated most of his thinking discussed above in this complex diagram. It is better to leave the readers to study this complex diagram than I am describing it verbose (Fig. 3).

Geddes has no concept of a smart city and IOT and ICT came much later. He understands how facts lead to thoughts and thoughts to action and then deeds. The instrumentality of reaching this is not there. There is no mention of how these concepts can be implemented through relevant institutions. There are no Governance Codes and Democracy Codes he put forward. He is credited with transforming physical planning with sociology and yet has not conceived the idea of the smart community then.

## 6 Lefebvre’s Spatial Triad and Differential Space for a Living [21, 22]

Urban space according to Lefebvre is a social construct. He extends the concept of Geddes further and what Doxiadis implied it in his work. He said, once you repeat social activities in the same place, it converts itself to a cultural construct. It is created through social relations that Lefebvre characterised as a triad of spaces: (1) spatial practice, (2) representations of space, and (3) spaces of representation [21]. Lefebvre is one of the few great twentieth century urban philosophers to engage directly with town planning, with a primary focus on urban regeneration [22, 23]. He was critical of the 1950s French new town of Mourenx, for the poor quality of urban public space created which he saw as suppressing democratic social interaction. Also, he was not impressed with the ‘regeneration’ plans for LesHalles; Paris’ wonderful nineteenth century produces markets. He related space with living that is the main reason in

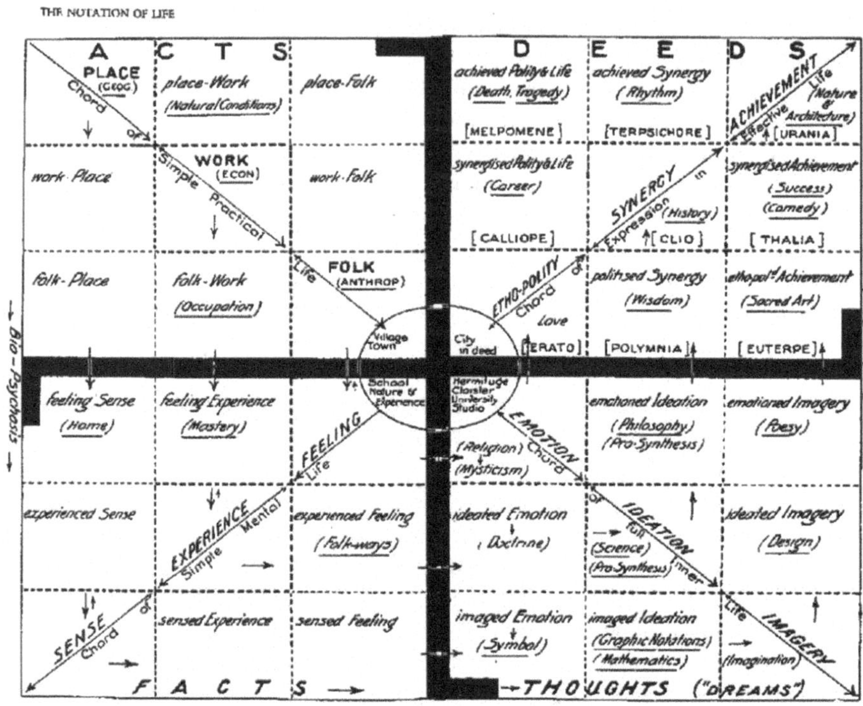


One of Geddes complex maps of a participatory holistic view of life (Image)

**Fig. 3** Geddes view of living

his contribution. Lefebvre believed that planners are the 'scientific' experts who create official representations of space, most notably, visual representations: zoning schemes, maps, plans, drawings, photographs, artistic impressions and policy documents. These representations are not neutral but imposed certain meanings onto urban space, imply how it should be used and not be used and by whom. A range of planning theories have emerged since the 1940s [24, 25] and there has been an emphasis on either substantive or procedural theory, both are important to living. Planning theory also contains elements of both positive and normative theory. If the theory is simply understanding in discipline and field of professional practice, then planners have adopted a few notable theoretical positions regarding the substantive and procedural nature of their subject over the decades. For hundreds of years, the formal planning of towns and cities was dominated by physical and architectural understandings of place and of practice. In the early twentieth century, such theoretical position coalesced in the writings of many theorists [24] albeit that these theorists had their own take on cities and planning. The physical planning approach also coalesced in two famous 'city beautiful' plans for North American cities: Chicago [25–28]. Similarly, in the UK high profile, post-World War Two reconstruction plans were made for Manchester, London and Glasgow. These theorists and practitioners focused on planning as ‘architecture writ large’, Daniel Burnham [29] saying famously ‘make no little plans’. They focused on the city as a physical ‘product’ constructed according to a precise blueprint that would lead to a desired physical end state. In the 1960s and 70 s, such ideas were challenged by those who understood planning, as social and community relations by Jacobs in 1960s, complex systems by McLoughlin (1969) [30] and Chadwick (1971) [31] and those who theorised planning as universal procedures or processes such as Faludi in 1973 [32]. From the 1990s considerable intellectual academic effort went into formulating understandings of planning around notions of who should be involved in the decision-making process and how: consequently, the so-called communicative or collaborative planning approach emerged [33–35].

Lefebvre’s production of space ideas remains contentious but highly relevant for the investigation of city-living- transformation in general and how the planning of urban space can contribute to social injustice. The spatial triad does have an intuitive simplicity and sees its elements—the spatial practice has three major elements: (1) the physical, material city and its routine maintenance; (2) major urban redevelopment in the context of existing neo-liberal capitalist and state power structures; (3) routines of daily life that conform with official representations of space. It is space directly perceptible through the senses, although mediated by 2 and 3 above—perceived space.

- **representations of space**: rational, intellectualised, official conceptions of urban areas for analytical, administrative and property development purposes. They are produced by technocrats: architects, engineers, urbanists and planners but also artists with a scientific bent. They are the dominant representations and may be in the form of the written word, for example in citywide zoning plans and strategy documents, or quasi-scientific visual representations of various kinds such as maps, master plans and design guides-conceived space.
- **spaces of representation**: has two major elements: (1) urban everyday space as directly lived by inhabitants and users in ways informed not so much by representations of space as by associated cultural memories, images and symbols imbued with cultural meaning; (2) emotional, artistic interpretations of city space by poets, writers and painters and others who create artistic representations of urban space. This kind of space overlays physical space and values places in ways that run counter to the dominant representations of space-imaginative and lived space.

Although Lefebvre refers to urban reality his ideas can be applied also to rural space, thereby establishing his habitat-oriented concept. In addition to the triad, two other Lefebvrian spatial concepts are important for the ways in which he understood urban space and theorised about cities, they are:

- **abstract space**; the urban spaces of state-regulated neo-capitalism characterised by their commodified exchange value and their tendency to homogenisation.
- **differential space**; privileges inclusiveness and use value rather than the exchange value of abstract space. It is often transitory space which can arise from the inherent vulnerabilities of abstract space.

What Lefebvre encouraged through the dissemination of ideas about urban space, was an understanding of the public space as not just an empty physical container but as a holistic product and social process where planners play a crucial role in producing representations that have a direct bearing on spatial practice (the built environment)and routine ways of life. His contribution to planning theory, therefore, is potentially profound and worthy of highlighting in this book. In seeking to redevelop urban public space and in abandoning urban space; public and private institutions it seems to produce inadvertently the potential for differential space. In tardy or rapid reactions to this potential, collective politicised action seizes opportunities to appropriate urban space for its use value, whether politically vibrant. Whether on a permanent basis running into years or on a temporary basis of months, days or hours, such appropriations are important for revealing the imminent vulnerabilities of abstract space. A theoretically robust critical appreciation of the production of urban space of use for planning requires a careful engagement with the histories of the elements of Lefebvre 's spatial triad. It will elevate diverse, inclusive urban public space theoretically, politically and practically to the status of one of the prime desired outcomes of planning practice. His main shortcoming was he did not include the ecological construct of a place in a city in his theory. How ecology can transform the spatial triads is left out in his theorising giving more attention to cultural than ecological. From the point of view of this book, he gave out the philosophy of the public realm. Smart living can only be developed through a variety of public realm meeting at the convenient location the psychological, social, cultural and biological needs of all living including man. Smart growth area can be a better location for the public realm and form-based codes standard can visualise the activities of the public realm which is the subject matter of Chap. 8 in this book.

## 7 Advaita Vedantic View of Living Through Awareness

Advaita (Non-dual) Vedantic (end of Veda) view of living is very briefly discussed by presenting, one Upanishad attributed to Mandukya [36–38] and Mandukya Karika which elaborate the Mandukya Upanishad by Gaudapada wrote in the seventh century AD, in a commentary known as Mandukya Karika. The Karika is based on Mandukya Upanishad that examines various states of the human being to determine the reality of existence as he experiences it which opens a way of living through awareness which Gaudapada calls as Asparsa Yoga without any dharmic codes of conduct. This awareness is also central to smart city concept as discussed earlier and Gaudapada put forward the concept of Asparsa Yoga in the concept of self-awareness to implement in daily living. Asparsais sparse, that is without touch. 'Sparsa' means touch, 'a' is used to denote opposite Sparsa. Mandukya Upanishad is one of the basic texts of Advaita Vedanta and Gaudapada advocates Asparsa Yoga to practice the Advaita way of lifeby. Apsara-yoga literally means 'yoga of non-contact' and therefore, point towards one of many ways of the practice of Hindu way of life or Advaita, non-dual living. Gaudapada in his Karika on Mandukya Upanishad, in Chapter IV, Aalati

Shanti, verse 2 speaks about Asparsa Yoga. This Asparsa Yoga is a word which is not used to anybody before Gaudapada. Gaudapada says that Brahma-jnani, the one who abides in the Self, is all-pervasive and there is nobody else other than him, to 'touch' or 'contact'. "I salute this Yoga, known as Asparsa Yoga, that is free from all touch, which implies duality, taught through scripture, the Yoga which promotes the happiness of all beings and conduces to the well-being of all, which is free from strife and contradictions." Verse 4 of Mandukya Karikastates: "The existent cannot pass into existence (birth). Nor can the non-existent to be born or come into being as existent. The Māndukya Karika has used the term 'Asparsa-yoga' only twice. Asparsa Yoga is the key soteriological (doctrine of salvation) term in the Mandukya Karika. Mentioned twice at prominent places, MK III. 39 and IV. 2, this term embodies the quintessence of Gaudapada's liberation philosophy (liberation from what?) and practice. In this work, this epithet has been especially used to denote the direct experience of the Ātman or the Self which is beyond the ken of all sense-organs and sense-experience. In this experience, the lower self and the consequent ego-sense gets into one leaving unalloyed bliss only. This is the same as 'jñāna.' Those who obtain the greatest happiness, the highest good and the absolute freedom is enjoyed by him free from all the doubts and misgiving. I bow down to that yoga known as Asparsa which is joyful to all beings, beneficial, free from dispute, non-contradictory, and set forth in the scriptures. —Mandukya Karika IV. 2

## *7.1 Mandukya Upanishad [36–38]*

Mandukya Upanishad is a 12-verse text as given below, probably a one-page document which describes the life and living in all details.

1. All is OM: The whole universe is the syllable, Om. Following is the exposition of Om. Everything that was is, or will be is, in truth Om. All else which transcends time, space, and causation is also Om.
2. Atman has Four Aspects: All of this, everywhere, is in truth Brahman, the Absolute Reality. This very Self itself, Atman, is also Brahman, the Absolute Reality. This Atman or Self has four aspects through which it operates.
3. First is Waking/Gross: The first aspect of Atman is the Self in the Waking state, *Vaishvanara*. In this first state, consciousness is turned outward to the external world. Through its seven instruments and nineteen channels, it experiences the gross objects of the phenomenal world.

The seven instruments are the more macrocosmic instruments, while the nineteen channels relate more to the microcosmic, individual person.

Seven Instruments: First, Consciousness manifests outward as elements space, air, fire, water, and earth, along with the individuation from the whole and the flow of energy (which we know as the pulsing impulse towards breath).

Nineteen Channels: Then, the individual operates through the four functions of mind (aspects of Antahkarana, the inner instrument), which are manas, chitta, ahamkara, and buddhi [39]. Those four operate through the five vayus (prana, Apana, Samana, Udana, and vyana), the five active senses or indriyas (karmendriyas of eliminating, procreating, moving, grasping, and speaking), and the five cognitive senses (jnanendriyas of smelling, tasting, seeing, touching, and hearing). Gaudpada who wrote a Karika for Mandukya Upanishad elaborate these again as follows.

- annamaya, the physical body which is the locus of action and enjoyment;
- pranamaya, the vital sheath which is manifest in the form of breath, the principle of animation and energy;
- manomaya, the mind which consists of desires;
- vijhdnamaya, having the power of cognition it is of the nature of the agent and is responsible for all empirical knowledge both in waking and dream states;
- anandamaya the mode of bliss.

Spirit and matter are thus the two extremes of man's being. Gaudapada accepts the Mandukyopanisad account of man's psycho-physical complex. He states Man, consists of:

- the five organs of perception (buddhindriyas) i.e., the organs of sight, sound, smell, taste and touch;
- the five organs of action {karmendriyas), i.e., the hands, feet, the organs of speech, generation, and evacuation;
- the five aspects of the vital breath [prana);
- the mind (manas);
- the intellect (buddhi);
- egoity (ahankara);
- the mind-stuff (Citta).

4. The Second is Dreaming/Subtle: The second aspect of Atman is the Self in the Dreaming state, Taijasa. In this second state, consciousness is turned towards the inner world. It also operates through seven instruments and nineteen channels discussed above, which engage the subtle objects of the mental realm (Fig. 4).
5. The third is Deep Sleep: The third aspect of Atman is the Self-operating in the Deep Sleep state, *Prajna*. In this third state, there is neither the desire for any gross or subtle object nor any dream sequences. In deep sleep, all such experiences have receded or merged into the ground of undifferentiated consciousness. Here, one is filled with the experience of bliss, and can also find the way to a clearer knowledge of the two preceding states.
6. Find the Experiencer: The one who experiences all these states of consciousness is the omniscient, indwelling source and director of all. This one is the womb out of which all the other emerges. All things originate from and dissolve back into this source.

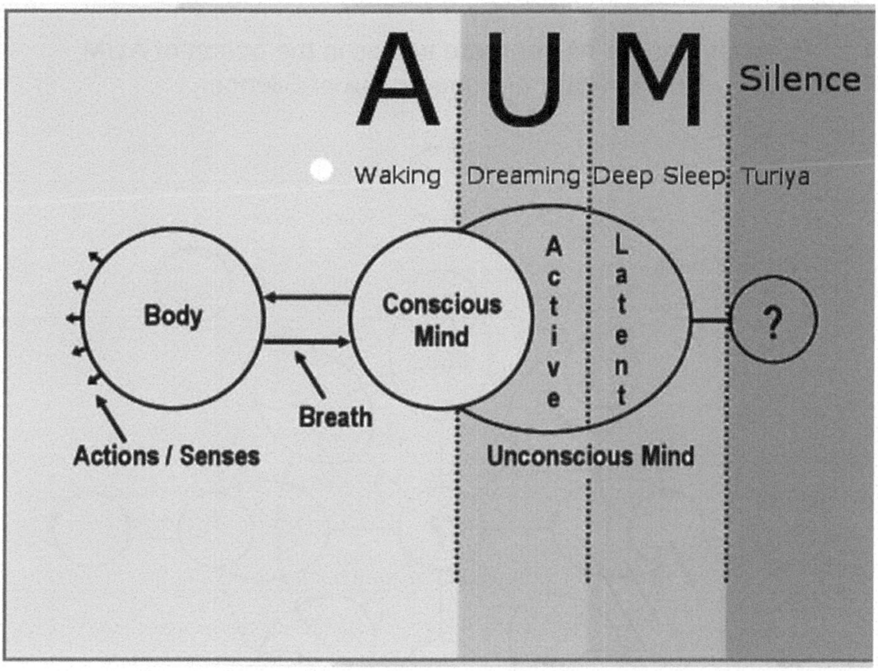


**Fig. 4** AUM. *Source* www.tripurashakti.com

7. The Fourth Aspect is Turiya: The fourth aspect of Atman or Self is Turiya, literally the fourth. In this fourth state, consciousness is neither turned outward nor inward. Nor is it both outward and inward; it is beyond both cognition and the absence of cognition. This fourth state of Turiya cannot be experienced through the senses or known, by comparison, deductive reasoning or inference; it is indescribable, incomprehensible, and unthinkable with the mind. This is Pure Consciousness itself. This is the real Self. It is within the cessation of all phenomena. It is serene, tranquil, filled with bliss, and is one without a second. This is the real or true Self that is to be realized (Fig. 5).
8. Those Four are the Same with "A-U-M" and Silence: That Om, though described as having four states, is indivisible; it is pure Consciousness itself. That Consciousness is Om. The three sounds A-U-M (ah, ou, mm) and the three letters A, U, M are identical with the three states of waking, dreaming, and sleeping, and these three states are identical with the three sounds and letters. The fourth state, *Turiya* is to be realized only in the silence behind or beyond the other three.
9. The Sound "A" is Waking/ Gross: *Vaishvanara* is the consciousness experienced during the waking state and is A, the first letter of Om. That simple sound of A is first and permeates all other sounds. One who is aware of this first level of reality has the fulfilment of all longings and is successful.

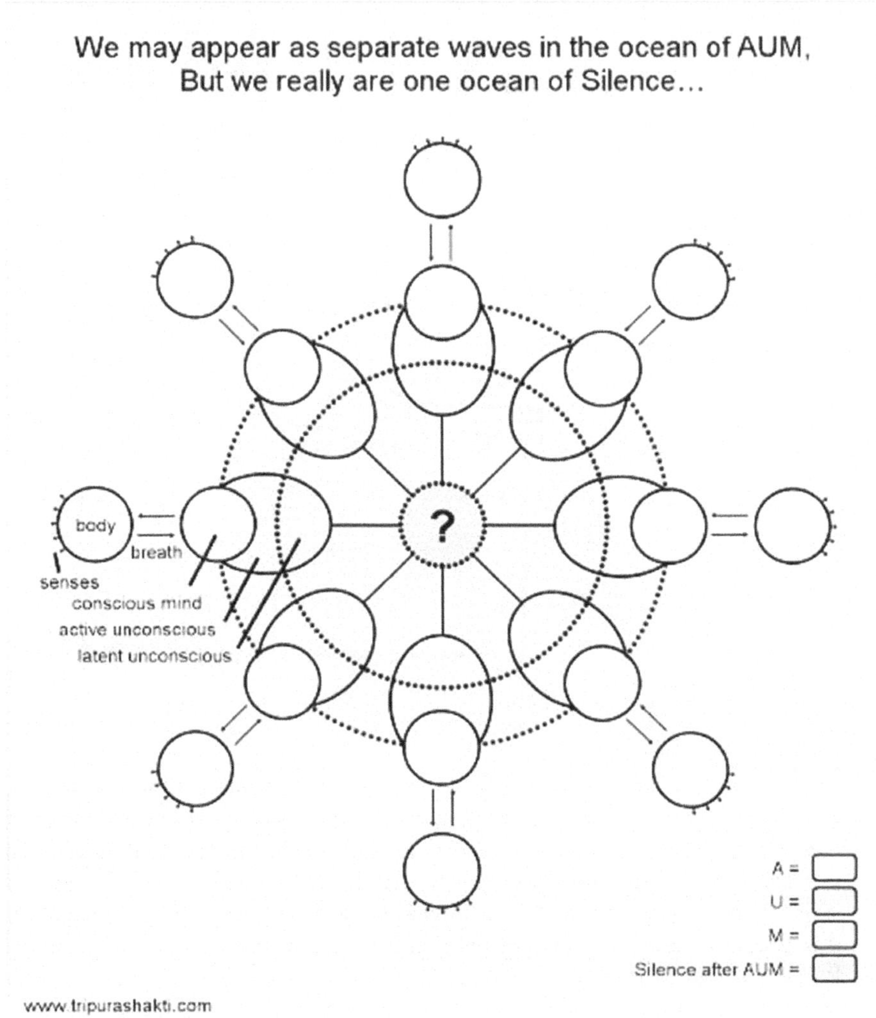


**Fig. 5** One Ocean of silence

10. The Sound "U" is Dreaming/ Subtle: *Taijasa* is the consciousness experienced during the dreaming state and is U, the second letter of Om. This intermediate state operates between the waking and sleeping states, reflecting some qualities of the other two. One who knows this subtler state is superior to others. For one who knows this, knowers of Brahman, the Absolute Reality, will be born into his family.

11. The Sound "M" is Deep Sleep/Causal: *Prajna* is the consciousness experienced during the state of a dreamless, deep sleep, and is M, the third letter of Om. It contains the other two and is that from which the other two emerge, and into the recede or merge. A knower of this subtler state can understand all within himself.
12. The silence after "A-U-M" is the True Self: The fourth aspect is the soundless aspect of Om. It is not speak-able and is not comprehended through the senses or by the mind. With the cessation of all phenomena, even of bliss, this soundless aspect becomes known. It is a state of nondual (Advaita) reality—one without a second. This fourth state, *Turiya*, is the real Self or true Self. One with direct experience of this expands to Universal Consciousness.

## 7.2 Asparsa Yoga a Brief Discussion Based on Gaudapada Karika [40–52]

A practitioner of Advaita Vedanta will not use the term yoga in his discussion for several reasons. Firstly, the strictly non-dualist perspective as stated by Gaudapada himself in MIC II. 32 and III. 36 precludes the very talk of "someone" doing "something" to attain "some goal." In MK III. 36, Gaudapada states that regarding the non-dual Brahman there can be no ceremony, conduct or practice (upacara). This, within the context of this book on Smart living for Smart cities; means this approach is without Governance codes and Democratic codes, without parliament and bureaucracy but only self-automated democracy using artificial intelligence based on the conviction that all are part of the total (Cultural and ecosystem). This can be achieved by Asparsa Yoga. Yet on an empirical level, it is obvious that the tradition does suggest that there are methods by which the realization of one's true nature can be facilitated or at least precipitated. Secondly, the term yoga itself denotes the coming into a union of two factors, one separated from the other. This is like the idea of sparsa suggested above and contrary to the concept of asparsa. The problem here is that the use of this kind of language is dualist oriented. Even to suggest that the yoga indicated here is the union of the individual soul or self (jiva) with the Supreme Soul (Brahman) alludes to the delusory belief in duality. Nonetheless, the term is used in a uniquely Advaita sense to refer metaphorically to union of the self with the self, not in the sense of gaining union with something separate, but rather in the realization of the unity of the Self as being without a second, i.e., as being the subtlest essence as well as the totality of non-dual reality. Further, there are numerous references to yoga within Advaita contexts of the Upanishads as well as in the literature of the developed tradition. The use of the term yoga in the sense of spiritual practice or discipline, in general, appears to be common to all the religion-philosophical traditions in India. Also, Gaudapada is teaching from within the milieu of yoga practitioners and many of his comments in this section are directly aimed at the erroneous views and practices of the various yoga disciplines. So, it is quite appropriate that he uses the term yoga in his discussions. Contextually, the term Asparsa Yoga refers to the practice or process by which the aspirant seeks to achieve this soteriological goal. The term is used in MK III. 39. This section comes after Gaudapada's refutation of the dualist views of reality and the establishment of the Advaita position. He begins the upadesa section after the mention of Maya and "mind creation" which sets the stage for a discussion of the nature of the mind and its relation to bondage and liberation (MK III. 29-38). This is specifically related to the notion of the state of "no-mind," which is given supreme valuation here by Gaudapada. At this point, the idea of Asparsa Yoga is introduced in MK III. 39 and is followed by a discussion of topics related to the actual practice of meditation. Thus, it appears that Asparsa Yoga refers to the nature of the process of liberation or Self-realization in this specifically Advaita Vedanta context.

## 7.3 Asparsa Yoga and a City Living from the Point of View of This Book

Upanishad was written in Sanskrit the language used by high caste Brahmins, the learned, and the elite of the later Vedic period but not the common man. Elite and Brahmins are well educated and intelligent and others were not. Such people do not require a procedural approach for Vedanta way of living or practice through Asparsa Yoga akin to Governance administered by bureaucracy for the ecological and cultural system of human settlement and Democracy Codes within the context of this smart city book. Buddha who targeted his teaching to all and used Pali as the language used by common persons had a contrary approach. He felt there is a need for a procedural approach to living to be spelt out with clarity and he implemented this approach which has high significance for the cultural and ecological system of human settlement for intervention. In Vedanta smart living is implicit but in Buddhism it is explicit. Buddha was influenced by many democratic republics that existed in his era and felt the need for Governance Codes and Democratic decision-making procedures in daily living which the author is more inclined to adopt.

This understanding, the AUM according to Vedanta school is one and only one that connects all living in a city to one non-divisible large smart community and give a way to view and interact with the cultural and ecological system that constitute the smart city without any governance code or democratic norms. Again, Mandukya Upanishad leaves all of you to decipher how to interact with the cultural and ecological system of a city with no dogmatic approach or procedural protocol and probably with current day artificial intelligence and data mining and not with man-made codes. No procedural theory of planning is given out in this short Upanishad purposively but Gaudapada showered light on its implementation in this Asparsa Yoga.

Asparsa yoga provides a deeper understanding how a person is totally integrated as part of cultural and ecosystem of a city in all his states waking, dreaming and deep sleeping state and beyond, and shows the way he partakes every second of his life as life presents the persons with itself in various forms in an unpredictable way. By understanding the totality of existence, he lives and acts with full understanding as an instinct on how to interact with the ecological and cultural system of a city without any help of Governing Codes, Democratic norms or even a smart community just like for example we do not codify the behaviour of eye for seeing and ear for hearing. Since the multi-dimensional environment that is sensed by various "mouths" as presented by Mandukya Upanishad of humans to experience can vary and be unpredictable the responses to these stimuli cannot be predetermined and codified in Advaita. A household who follows the path of Asparsa Yoga shall start integrating with the cultural and ecological system all his life spontaneously with the act of living by Asparsa Yoga just like a cat thrown into water start swimming without swimming lessons/swimming codes.

The Māndūkya is the essence of all the Upanishads, a study and assimilation of which, alone, is enough to lead one to emancipation. For the liberation of the seeker, the Māndūkya Upanishad, alone, is adequate, if it is properly digested into the experience. You absorb it into your minds and make it a part of your practical life. "He is a real brāhmaṇa who knows this secret, who has this knowledge, who lives this knowledge, and to whom this knowledge is practice, to whom action is not different from having this knowledge where karma and jñāna come together in a fraternal embrace, where there is no friction between work and contemplation, where life becomes realisation, where work becomes worship and God-consciousness, where one's very existence becomes a blessedness to all earth, where one's life on earth becomes a teaching, where example becomes a precept, and where one becomes a representative of Īsvara in this world". This is the gist of the Māndūkya Upanishad". This knowledge is greater than the wealth of all the world, a saviour of humanity from the difficulties of a finite life, a direct means to moksha, Immortal Existence, the great Goal of your lives.

The vision of smart living postulated by Advaita, in short, have no Governing Code, no Democratic ethical norms, no IOT and ICT but artificial intelligence emanating with the idea of all is part of ecological and cultural system drives smart living.

## 8 The Early Buddhist View of Living [53–55]

Over 2500 years ago at the end of the Vedic period, the Buddha proclaimed an eightfold path for ending suffering out of many issues of conflicts and contradictions we experience in this cultural and ecological system where we live. The Eightfold Path was announced at the enlightenment of the Buddha as the way to end suffering. It was preceded by the 4 Noble Truths. As against Asparsa Yoga Buddhism was meant for a common man and not an all understanding and integrated self which Asparsa Yoga assumes or give rise to but common men who require Governing and Democratic norms and codes and a smart community to conduct living with the ecological and cultural system. The reason behind this approach was Buddha greatly appreciate the democratic system of Governance in existence in many Republics at that time in many parts of India. This ancient Republican culture Buddhist practice adopted in Buddhist Sangha in many countries can be evolved into the smart community of today to implement smart living with ICT and IOT additions.

### *8.1 The Four Noble Truth [56, 57]*

1. The Truth of *Dukkha*: Existence in all its manifestation is suffering because it is in a state of constant change. Constant changes stress the cultural and ecological system from time to time. Everything in life is impermanent, according to the Buddha, and is therefore unattainable in any real sense, and this causes suffering or *dukkha*. There are many types of *dukkha* or suffering. The wheel of *samsara* or constant *reincarnation* is dukkha; birth, ageing, illness and death are *dukkha*; sorrow, lamentation, pain, grief and despair are *dukkha*; not getting what you want is *dukkha*. The understanding of impermanence is *dukkha*. Even enjoying pleasures in life is *dukkha because* you know they will eventually fade or go away. This simple statement appeals to the common man.
2. The Truth of the Origin of *Dukkha:* Attachment to pleasure and aversion to pain trap you within existence. If you stay in the up-down, of the dance of existence, you're stuck. You lunge with craving fingers towards the things you want, and you recoil in horror at the things that you do not want. Or you just start fidgeting and flailing in your existence. And existence is like quicksand and sucks you in deeper the more you struggle. You are not still, and so you cannot see the truth. These tendencies of the mind are called the *kleshas*. Roughly translated, they are Attachment, Aversion, Ignorance, Ego and Clinging to Life.
3. The Truth of Cessation of *Dukkha*: By ceasing the attachment/aversion cycle, you can experience liberation from existence. Attachment and Aversion are the fundamental movements of the mind. Cease the craving that drives them, and you will cease the motion of the mind. Renounce the sense pleasures of the world, and Just an existence no longer have a hold on you.
4. The Truth of the Path to the Cessation of Dukkha: The 8-Fold Path is the way to end suffering. The 4 Noble Truths were the Buddha's basic diagnosis of humanity's ailment: We are entangled in our life produced by our own attachments and aversions and are unhappy because we have forgotten that it's just a game to be appreciated in proper perspective, about the inherent losses and gains in a game. The Buddha prescribes Eight Steps to deal with this situation.

These constitute the Noble Eightfold Path, as follows:

## 8.2 The Eight-Fold Path [58, 59]

1. Right View: Right View means understanding that life is just a game. It is transitory, and it causes suffering if you think it is real and takes it too seriously.
2. Right Intention: Right Intention means being resolved to renunciation. It doesn't necessarily mean that you give up everything and go become a monk. It is working to steadily improve yourself, to rid yourself of the defilements of the world. It means intending good will and harmlessness towards all beings (including yourself) instead of ill will and harmfulness.
3. Right Speech: Right Speech means not lying, not speaking divisively or abusively, and not engaging in idle chatter and gossip. Words are magic. They create the reality you live in. Be mindful of how you use them. Do not waste them.
4. Right Action: Right Action means being morally upright in your actions, and not acting in corrupt or harmful ways. Specifically, it means not killing, not stealing that is not yours, and not engaging in sexual misconduct.
5. Right Livelihood: Right Livelihood is not making your living in a way that creates suffering for other beings. There are five specific types of business that should not be undertaken:

    - Business in Weapons. (Arms dealing.)
    - Business in Humans. (Slavery, prostitution, human trafficking, prisons-for-profit.)
    - Business in Meat. (Selling meat or raising animals for slaughter.)
    - Business in Intoxicants. (Dealing, selling or advertising drugs, alcohol or tobacco.)
    - Business in Poison. (Dealing, selling or advertising any chemical designed to cause harm.)

Ideally, your business should be something you can put your heart into, and that benefits other sentient beings.

6. Right Effort: Right Effort means working to uphold all that is wholesome to yourself and the world and working to abandon all that is harmful.
7. Right Mindfulness: Right Mindfulness means staying present and aware of yourself, reality and your place. It means paying attention. You can easily take harmful actions without mindful.
8. Right Concentration: Right Concentration means meditation. Once free from sense-desires, and leading a tranquil life, the practitioner begins to meditate, usually through mindfulness of breathing, object concentration or repetition of mantras. The meditator may begin to pass through the four classic Jnanas or meditative absorption, being:

- The rapture and joy of withdrawal into meditation;
- The rapture and joy of unified awareness, free from directed thought;
- Equanimity, mindfulness, and alertness (as rapture fades away);
- The perfection of equanimity and mindfulness, free from both pleasure and pain. Pure, bright awareness.

The Buddhist way of life accepts the cultural and ecological system in a city that envelops all. It understands the influence of this system on human existence and prescribes in concrete terms understandable by a common man how to deal with it and conduct living. Buddhism cannot be there without the Sangha, the modern-day smart community and cannot be there without the Governing codes and Democratic norms of the much appreciated and celebrated Republics in the Buddhist era.

## *8.3 Early Buddhism and Smart Living Knowledge Base*

Smart aspect of smart living is a very new development and it was not there in the early Buddhist view of living. Next to important in the concept is a smart community, again smart is not there but the community is advocated in Sangha which practised Democracy in decision making was mostly influenced by a Republican form of Government practised at that time. Governance was very much there in Buddhism which gave specific rules of Governance which is discussed below of high significance in living.

The eight-fold path was what Buddha preached to overcome four noble truths. It contains moral content the essential of governing principle for living, mental discipline and wisdom which includes ecological wisdom. His approach was to explain all these in a way people can imbibe in their lives depending on the level of development of people who listen to his discourses for 45 years.

Ethical content is Buddhas teaching on universal love and compassion for all living being. His teaching was there for good for many, happiness for many, out of compassion for the world, which I consider the ecological principle of living. It is based on his teaching of wisdom and compassion two quality that should be developed for a living. Right speech, right action, and right livelihood in his 8-fold path advocate mental discipline in the right effort, right mindfulness and right concentration. The remaining two right thought and right understanding constitute wisdom for living. Smart living in the smart city involves continuous education and training for smart community and the subject matter of the Buddhist way of living on ethical compassion for all living, mental discipline through meditation and practice of wisdom.

## 9 Smart Living Study Focus for Secular Countries

Secular countries like India cannot give more important politically to anyone religion even if it is followed by a majority. All religions shall exist side by side with equal state patronage, respect and a smart living system can be built on this basis such that in areas that are classified into areas where one religion predominates and then design an appropriate smart living system around this area with this religion. Legislation at Central and federal level can be to meet the special needs of these religious group. This results in a heterogeneous smart living system spatially segregated within the same metropolises since you can find such areas and pockets in any cities. Local community-level democracy for decision making can be built around the predominant religion. For example, Zakat a practice of compulsory charity by Muslims can be subjected to the democratic decision making of the community in that area where Islam predominates.

Another approach which is applicable both for secular and to theocratic countries is to focus on the comparative descriptive statistics of the states and identify a unique positive characteristic of the study area and develop smart living centred around the strong positive indicators than the negative and weakness indicators. The positive strong indicators have come into existence in cooperation with the cultural and ecological system supporting it than negative and weak. We have done this exercise for one study detailed out in Chapter "Geospatial Intelligence for Smart Living–Case of New Delhi". The rationale for selecting the indicator for this study is presented briefly with a few statistics presented.

In the state of Kerala where one study area falls, we found comparatively the achievement in health was better than other states and took smart living research based on health. We present the table of time series data on the life expectancy of states in India and found Kerala states lead in it. Behind this phenomenon, there have been actions in many spheres of life in Kerala for centuries. These systems still exist in terms of socio-cultural institutions that can be used to design the smart living system of the study area in Kerala. The State wise life expectancy at birth and infant mortality comparisons are presented below (Table 2).

Life Expectancy at Birth by Gender, 1992–2004 Archived 2016-03-05 at the Way back Machine. Table 5A.10, India Human Development Index Report, page 287.

State wise life expectancy at birth over the years

| States | 1981–1991 | 1988–1992 | 1989–1993 | 1998–2002 |
|---|---|---|---|---|
| Andhra Pradesh | 59.9 | 60.2 | 60.6 | 63.5 |
| Assam | 53.9 | 54.1 | 54.9 | 57.9 |
| Bihar | 56.7 | 57.5 | 58.5 | 60.8 |
| Gujarat | 58.8 | 59.5 | 60.1 | 63.4 |
| Haryana | 62.7 | 62.5 | 62.9 | 65.5 |
| Karnataka | 61.7 | 62.2 | 61.9 | 64.5 |

(continued)

(continued)

| | | | | |
|---|---|---|---|---|
| **Kerala** | **70.4** | **71.3** | **72** | **73.5** |
| Madhya Pradesh | 53.3 | 53.4 | 54 | 56.9 |
| Maharashtra | 63 | 63.4 | 64.2 | 66.2 |
| Orissa | 55 | 55.4 | 55.5 | 58.5 |
| Punjab | 66.4 | 66.6 | 66.4 | 68.5 |
| Rajastan | 55.8 | 56.3 | 58 | 61.1 |
| Tamil Nadu | 61.1 | 61.5 | 62.4 | 65.2 |
| Uttar Pradesh | 54.7 | 55.4 | 55.9 | 59.1 |
| W.B. | 61.2 | 61.4 | 61.5 | 63.9 |
| All India | 58.2 | 58.7 | 59.4 | 62.5 |

*Source* (i) Selected Socio Economic Statistics, India 1996–97
(ii) Economic Survey 2005–2006, p-SI 11

Here the state of Kerala is in a leading position. We also looked at a few more indicators to substantiate our claim. We looked at the Infant mortality time series data of the state. This is presented below and seems to follow the life expectancy of the state giving the leading position to a state of Kerala (Table 3).

Infant mortality rate at national and state levels

| States | 1975 | 1980 | 1985 | 1990 | 1995 | 1997 | 1998 | 2003 |
|---|---|---|---|---|---|---|---|---|
| Andhra Pradesh | 123 | 92 | 83 | 70 | 63 | 63 | 66 | 59 |
| Assam | 144 | 103 | 111 | 76 | 76 | 76 | 78 | 67 |
| Bihar | – | – | 106 | 75 | 73 | 71 | 67 | 60 |
| Gujarat | 154 | 113 | 98 | 72 | 62 | 62 | 64 | 57 |
| Haryana | 114 | 103 | 85 | 69 | 69 | 68 | 69 | 59 |
| Karnataka | 80 | 71 | 69 | 70 | 62 | 53 | 58 | 52 |
| **Kerala** | **54** | **40** | **31** | **17** | **15** | **12** | **16** | **11** |
| Madhya Pradesh | 151 | 142 | 122 | 11 | 99 | 94 | 97 | 82 |
| Maharashtra | 92 | 75 | 68 | 58 | 55 | 47 | 49 | 42 |
| Orissa | 149 | 143 | 132 | 122 | 103 | 96 | 98 | 83 |
| Punjab | 98 | 89 | 71 | 61 | 54 | 51 | 54 | 49 |
| Rajastan | 155 | 105 | 108 | 84 | 86 | 85 | 83 | 75 |
| Tamil Nadu | 112 | 93 | 81 | 59 | 54 | 53 | 53 | 43 |
| Uttar Pradesh | 198 | 159 | 142 | 99 | 86 | 85 | 85 | 76 |
| West Bengal | N A | N A | 74 | 63 | 58 | 55 | 53 | 46 |
| All India | 140 | 114 | 97 | 80 | 74 | 71 | 72 | 60 |

*Source* Sample Registration System, Government of India, 2003

We also found that the State of Kerala has many more issues connected with health. A time series of data of morbidity is given below (Table 4).

**Table 2** Time series life expectancy by birth of states in India

| Rank | State | Life expectancy at birth | |
|---|---|---|---|
| | | (2010–14) | (2002–06) |
| 1 | Kerala | 74.9 | 74.0 |
| 2 | Delhi | 73.2 | – |
| 3 | Jammu and Kashmir | 72.6 | – |
| 4 | Uttarakhand | 71.7 | 60.0 |
| 5 | Himachal Pradesh | 71.6 | 67.0 |
| 5 | Punjab | 71.6 | 69.4 |
| 5 | Maharashtra | 71.6 | 67.2 |
| 8 | Tamil Nadu | 70.6 | 66.2 |
| 9 | West Bengal | 70.2 | 64.9 |
| 10 | Karnataka | 68.8 | 65.3 |
| 11 | Gujarat | 68.7 | 64.1 |
| 12 | Haryana | 68.6 | 66.2 |
| 13 | Andhra Pradesh (includes Telangana) | 68.5 | 64.4 |
| 14 | Bihar | 68.1 | 61.6 |
| * | **India** | **67.9** | **63.5** |
| 15 | Rajasthan | 67.7 | 62.0 |
| 16 | Jharkhand | 66.6 | 58.0 |
| 17 | Odisha | 65.8 | 59.6 |
| 18 | Chattisgarh | 64.8 | 58.0 |
| 19 | Madhya Pradesh | 64.2 | 58.0 |
| 20 | Uttar Pradesh | 64.1 | 60.0 |
| 21 | Assam | 63.9 | 58.9 |

*Source* "Abridged life tables 2010–14" (PDF). Office of the Registrar General & Census Commissioner, India, p. 5. Retrieved 12 January 2017

Also, we looked at the anaemic state of women across states which is not showing Kerala state is high performing.

State-wise prevalence of anaemia among Women and Children in India (1998–1999)

| States/UTs | Percent of Women With | | | | |
|---|---|---|---|---|---|
| | Mild Anaemia | Moderate Anaemia | Severe Amaemia | Percent of Women Age 15–49 with Any Anaemia | Age 6–35 Months with Any Anaemia |
| Andhra Pradesh | 32.5 | 14.9 | 2.4 | 49.8 | 72.3 |

(continued)

(continued)

| States/UTs | Percent of Women With | | | | |
|---|---|---|---|---|---|
| | Mild Anaemia | Moderate Anaemia | Severe Amaemia | Percent of Women Age 15–49 with Any Anaemia | Age 6–35 Months with Any Anaemia |
| Assam | 43.2 | 25.6 | 0.9 | 69.7 | 63.2 |
| Bihar* | 42.9 | 19 | 1.5 | 63.4 | 81.3 |
| Gujarat | 29.5 | 14.4 | 2.5 | 46.3 | 74.5 |
| Haryana | 30.9 | 14.5 | t.6 | 47 | 83.9 |
| Karnataka | 26.7 | 13.4 | 2.3 | 42.4 | 70.6 |
| Kerala | 19.5 | 2.7 | 0.5 | 22.7 | 43.9 |
| Madhya Pradesh* | 37.6 | 15.6 | 1 | 54.3 | 75 |
| Maharashtra | 31.5 | 14.1 | 2.9 | 48.5 | 76 |
| Orissa | 45.1 | 16.4 | 1.6 | 63 | 72.3 |
| Punjab | 28.4 | 12.3 | 0.7 | 41.4 | 80 |
| Rajasthan | 32.3 | 14.1 | 2.1 | 48.5 | 82.3 |
| Tamil Nadu | 36.7 | 15.9 | 3.9 | 56.5 | 69 |
| Uttar Pradesh* | 33.5 | 13.7 | 1.5 | 48.7 | 73.9 |
| West Bengal | 45.3 | 15.9 | 1.5 | 62.7 | 78.3 |
| India | 35 | 14.8 | 1.9 | 51.8 | 74.3 |

*Source* Women and Men in India 2004, Ministry of Statistics and Programme Implementation, Govt, of India & India Yearbook 2003, Manpower Profile

A look at the morbidity pattern as presented in the table below also shows the same conclusion as before (Table 5).

The incidence of Cancer is higher than all India figures for Kerala (Fig. 6).

Also presented below is hypertension statistic age-standardised prevalence of hypertension for above 18 years old (Fig. 7).

Death due to coronary heart disease is given below (Figs. 8, 9, and 10).

Another factor emerged looking at the district figures within Kerala. Health indices in the districts of Kerala are presented below. There are wide variations between districts and therefore the smart living system can be designed for districts and metropolitan agglomeration and not the state level.

**Table 3** Infant Mortality Rate (IMR) (per 1000 live berths)

| Total (Male + Female) | | | | | | | | | | | | | | | | | |
|---|---|---|---|---|---|---|---|---|---|---|---|---|---|---|---|---|---|
| State/UT | 2016 | 2015 | 2014 | 2013 | 2012 | 2011 | 2010 | 2009 | 2008 | 2007 | 2006 | 2005 | 2004 | 2003 | 2002 | 2001 | 2000 |
| Andhra Pradesh | 34 | 37 | 39 | 39 | 41 | 43 | 46 | 49 | 52 | 54 | 56 | 57 | 59 | 59 | 62 | 66 | 65 |
| Arunachal Pradesh | 36 | 30 | 30 | 24 | 24 | 23 | 25 | 27 | 31 | 31 | 31 | 27 | 25 | 21 | 23 | 7 | 16 |
| Assam | 44 | 47 | 49 | 54 | 55 | 55 | 58 | 61 | 64 | 66 | 67 | 68 | 66 | 67 | 70 | 74 | 75 |
| Bihar | 38 | 42 | 42 | 42 | 43 | 44 | 48 | 52 | 56 | 58 | 60 | 61 | 61 | 60 | 61 | 62 | 62 |
| Chattisgarh | 39 | 41 | 43 | 46 | 47 | 48 | 51 | 54 | 57 | 59 | 61 | 63 | 60 | 70 | 73 | 77 | 79 |
| Delhi | 18 | 18 | 20 | 24 | 25 | 28 | 30 | 33 | 35 | 36 | 35 | 35 | 32 | 26 | 33 | 25 | 32 |
| Goa | 8 | 9 | 10 | 9 | 10 | 11 | 10 | 10 | 10 | 10 | 15 | 16 | 18 | 13 | 16 | 14 | 21 |
| Gujarat | 30 | 33 | 35 | 36 | 38 | 41 | 44 | 48 | 50 | 52 | 53 | 54 | 53 | 57 | 60 | 60 | 62 |
| Haryana | 33 | 36 | 36 | 41 | 42 | 44 | 48 | 51 | 54 | 55 | 57 | 60 | 61 | 59 | 62 | 66 | 67 |
| Himachal Pradesh | 25 | 28 | 32 | 35 | 36 | 38 | 40 | 45 | 44 | 47 | 50 | 49 | 51 | 42 | 61 | 43 | 51 |
| Jammu & Kashmir | 24 | 26 | 34 | 37 | 39 | 41 | 43 | 45 | 49 | 51 | 52 | 50 | 49 | 44 | 47 | 48 | 50 |
| Jharkhand | 29 | 32 | 34 | 37 | 38 | 37 | 42 | 44 | 46 | 48 | 49 | 50 | 49 | 51 | 58 | 62 | 70 |
| Karnataka | 24 | 28 | 29 | 31 | 32 | 35 | 38 | 41 | 45 | 47 | 48 | 50 | 49 | 52 | 55 | 58 | 57 |
| Kerala | 10 | 12 | 12 | 12 | 12 | 12 | 13 | 12 | 12 | 13 | 15 | 14 | 12 | 11 | 10 | 11 | 14 |
| Madhya Pradesh | 47 | 50 | 52 | 54 | 56 | 59 | 62 | 67 | 70 | 72 | 74 | 76 | 79 | 82 | 85 | 86 | 88 |
| Maharashtra | 19 | 21 | 22 | 24 | 25 | 25 | 28 | 31 | 33 | 34 | 35 | 36 | 36 | 42 | 45 | 45 | 48 |
| Manipur | 11 | 9 | 11 | 10 | 10 | 11 | 14 | 16 | 14 | 10 | | | | | | 10 | 22 |
| Meghalaya | 39 | 42 | 46 | 47 | 49 | 52 | 55 | 59 | 58 | 52 | 53 | 49 | 43 | 56 | 66 | 51 | 66 |
| Mizoram | 27 | 32 | 32 | 35 | 35 | 34 | 37 | 36 | 37 | | | | 27 | 16 | 6 | 18 | 18 |
| Nagaland | 12 | 12 | 14 | 18 | 18 | 21 | 23 | 26 | 26 | 19 | 20 | 18 | 17 | NA | NA | NA | NA |

(continued)

**Table 3** (continued)

| Total (Male + Female) | | | | | | | | | | | | | | | | | |
|---|---|---|---|---|---|---|---|---|---|---|---|---|---|---|---|---|---|
| State/UT | 2016 | 2015 | 2014 | 2013 | 2012 | 2011 | 2010 | 2009 | 2008 | 2007 | 2006 | 2005 | 2004 | 2003 | 2002 | 2001 | 2000 |
| Odisha | 44 | 46 | 49 | 51 | 53 | 57 | 61 | 65 | 69 | 71 | 73 | 75 | 77 | 83 | 87 | 91 | 95 |
| Punjab | 21 | 23 | 24 | 26 | 28 | 30 | 34 | 38 | 41 | 43 | 44 | 44 | 45 | 49 | 51 | 52 | 52 |
| Rajasthan | 41 | 43 | 46 | 47 | 49 | 52 | 55 | 59 | 63 | 65 | 67 | 68 | 67 | 75 | 78 | 80 | 79 |
| Sikkim | 16 | 18 | 19 | 22 | 24 | 26 | 30 | 34 | 33 | 33 | 33 | 30 | 30 | 22 | 25 | 29 | 47 |
| Tamil Nadu | 17 | 19 | 20 | 21 | 21 | 22 | 24 | 28 | 31 | 35 | 37 | 37 | 41 | 43 | 44 | 49 | 51 |
| Telangana | 31 | 34 | | | | | | | | | | | | | | | |
| Tripura | 24 | 20 | 21 | 26 | 28 | 29 | 27 | 31 | 34 | 41 | 36 | 31 | 30 | 23 | 33 | 35 | 35 |
| Uttar Pradesh | 43 | 46 | 48 | 50 | 53 | 57 | 61 | 63 | 67 | 69 | 71 | 73 | 72 | 76 | 80 | 83 | 83 |
| Uttarakhand | 38 | 34 | 33 | 32 | 34 | 36 | 38 | 41 | 44 | 48 | 43 | 42 | 42 | 41 | 44 | 48 | 50 |
| West Bengal | 25 | 26 | 28 | 31 | 32 | 32 | 31 | 33 | 35 | 37 | 38 | 38 | 40 | 46 | 49 | 51 | 51 |
| Andaman & Nicobar Islands | 16 | 20 | 22 | 24 | 24 | 23 | 25 | 27 | 31 | 31 | 31 | 27 | 25 | 21 | 23 | 7 | 16 |
| Chandigarh | 14 | 21 | 23 | 21 | 20 | 20 | 22 | 25 | 28 | 35 | 23 | 19 | 22 | 12 | 22 | 16 | 27 |
| D&N Haveli | 17 | 21 | 26 | 31 | 33 | 35 | 38 | 37 | 34 | 29 | 24 | 43 | 34 | 49 | 51 | 61 | 57 |
| Daman & Diu | 19 | 18 | 18 | 20 | 22 | 22 | 23 | 24 | 31 | 28 | 37 | 28 | 30 | 36 | 30 | 43 | 54 |
| Lakshadweep | 19 | 20 | 20 | 24 | 24 | 24 | 25 | 25 | 31 | 38 | 25 | 22 | 39 | 21 | 15 | 38 | 22 |
| Puducherry | 10 | 11 | 14 | 17 | 17 | 19 | 22 | 22 | 25 | 16 | 28 | 28 | 26 | 27 | 25 | 22 | 20 |
| India | 34 | 37 | 39 | 40 | 42 | 44 | 47 | 50 | 53 | 55 | 57 | 58 | 58 | 60 | 63 | 66 | 68 |

*Source* Sample Registration System
*Source URL* http://niti.gov.in/content/infant-mortality-rate-imr-1000-live-births
*Source* http://niti.gov.in

**Table 4** State wise morbidity rates in India (1998–99)

| State | Number of persons per 1000 suffering from | | | | Case fatality rate per 1000 cases tuberculosis |
|---|---|---|---|---|---|
| | Asthma | Tuberculosis | Jaundice | Malaria | |
| Andhra Pradesh | 42.92 | 5.92 | 15.71 | 48.51 | 8.32 |
| Assam | 32.78 | 7.10 | 27.68 | 29.74 | NA |
| Bihar | 20.28 | 9.89 | 15.15 | 37.88 | NA |
| Gujarat | 19.79 | 4.35 | 11.09 | 44.49 | 2.33 |
| Haryana | 19.22 | 3.58 | 9.93 | 20.93 | 7.37 |
| Himachal Pradesh | 13.39 | 2.59 | 4.50 | 3.74 | 18.22 |
| Jammu and Kashmir | 17.25 | 4.28 | 9.42 | 6.02 | 0.00 |
| Karnataka | 17.33 | 2.69 | 3.73 | 6.00 | 14.28 |
| **Kerala** | **48.06** | **5.26** | **5.28** | **0.56** | **5.39** |
| Madhya Pradesh | 22.73 | 6.02 | 19.27 | 100.15 | 1.60 |
| Maharashtra | 25.24 | 2.82 | 15.34 | 40.98 | 8.50 |
| Orissa | 32.55 | 8.33 | 12.53 | 74.14 | 9.59 |
| Punjab | 13.08 | 2.07 | 9.76 | 10.82 | 4.76 |
| Rajasthan | 30.73 | 3.97 | 10.05 | 40.99 | 5.24 |
| Tamil Nadu | 15.46 | 4.79 | 11.42 | 3.80 | 5.91 |
| Uttar Pradesh | 19.79 | 5.51 | 9.63 | 35.52 | 4.27 |
| West Bengal | 25.93 | 4.92 | 23.81 | 14.82 | 25.57 |
| India | 24.68 | 5.44 | 13.61 | 36.97 | 9.04 |

*Source* NFHS-II and Health Information of India, 1999

**Table 5** State wise morbidity rates in India (1998–99)

| State | Number of persons per 1000 suffering from | | | | Case fatality rate per 1000 cases tuberculosis |
|---|---|---|---|---|---|
| | Asthma | Tuberculosis | Jaundice | Malaria | |
| Andhra Pradesh | 42.92 | 5.92 | 15.71 | 48.51 | 8.32 |
| Assam | 32.78 | 7.10 | 27.68 | 29.74 | NA |
| Bihar | 20.28 | 9.89 | 15.15 | 37.88 | NA |
| Gujrat | 19.79 | 4.35 | 11.09 | 44.49 | 2.33 |
| Haryama | 19.22 | 3.58 | 9.93 | 20.93 | 7.37 |
| Himachal Pradesh | 13.39 | 2.59 | 4.50 | 3.74 | 18.22 |
| Jammu and Kashmir | 17.25 | 4.28 | 9.42 | 6.02 | 0.00 |

(continued)

**Table 5** (continued)

| State | Number of persons per 1000 suffering from | | | | Case fatality rate per 1000 cases tuberculosis |
|---|---|---|---|---|---|
| | Asthma | Tuberculosis | Jaundice | Malaria | |
| Karnataka | 17.33 | 2.69 | 3.73 | 6.00 | 14.28 |
| **Kerala** | **48.06** | **5.26** | **5.28** | **0.56** | **5.39** |
| Madhya Pradesh | 22.73 | 6.02 | 19.27 | 100.15 | 1.60 |
| Maharashtra | 25.24 | 2.82 | 15.34 | 40.98 | 8.50 |
| Orissa | 32.55 | 8.33 | 12.53 | 74.14 | 9.59 |
| Punjab | 13.08 | 2.07 | 9.76 | 10.82 | 4.76 |
| Rajasthan | 30.73 | 3.97 | 10.05 | 40.99 | 5.24 |
| Tamil Nadu | 15.46 | 4.79 | 11.42 | 3.80 | 5.91 |
| Uttar Pradesh | 19.79 | 5.51 | 9.63 | 35.52 | 4.27 |
| West Bengal | 25.93 | 4.92 | 23.81 | 14.82 | 25.57 |
| India | 24.68 | 5.44 | 13.61 | 36.97 | 9.04 |

*Source* NFHS-II and Health Information of India, 1999

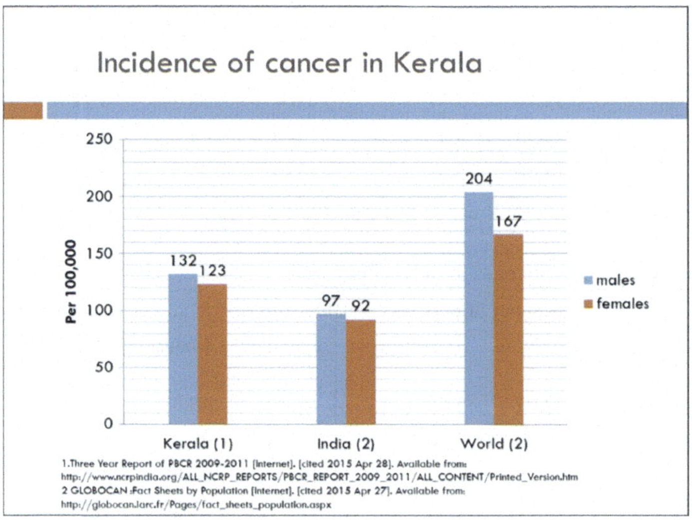


**Fig. 6** Incidence of cancer in Kerala

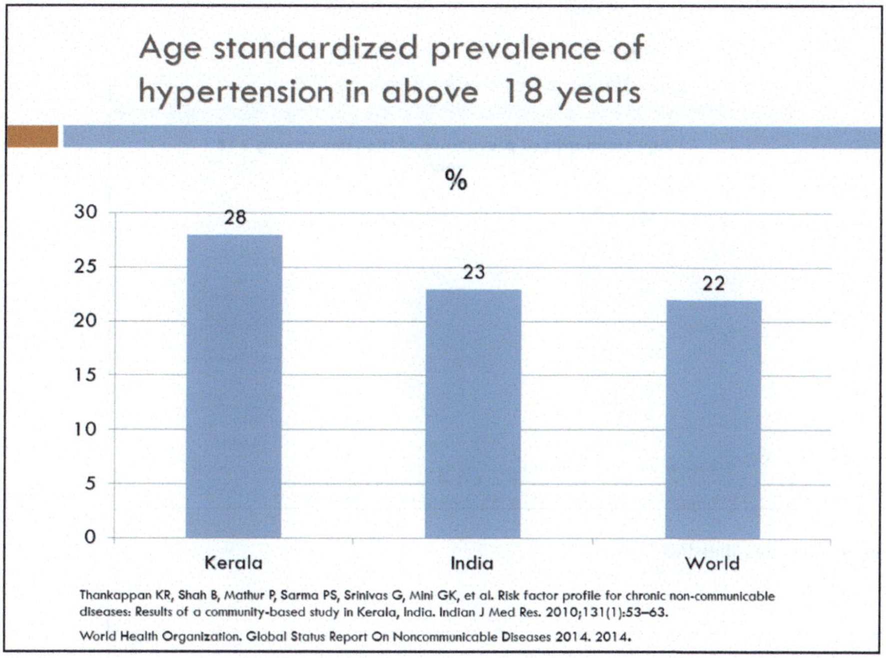


**Fig. 7** Age-standardised prevalence of hypertension in the above 18 years

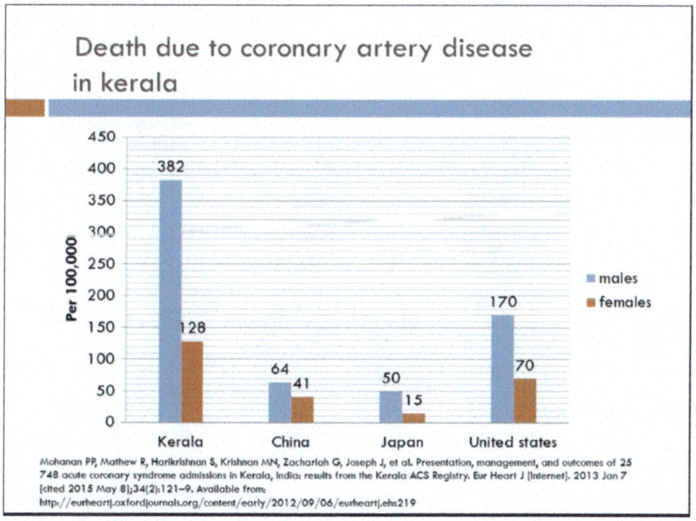


**Fig. 8** Death due to coronary artery diseases in Kerala

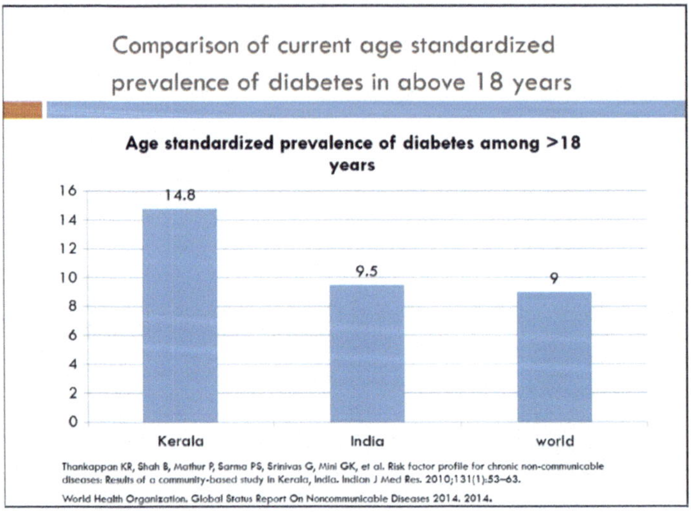


**Fig. 9** Diabetes statistics

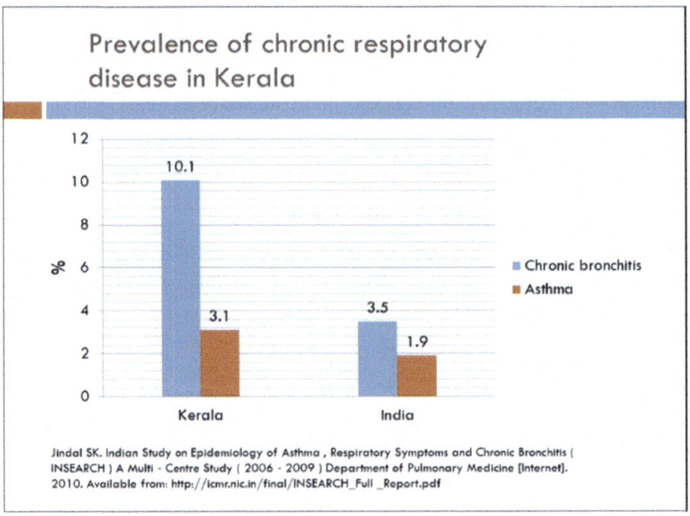


**Fig. 10** Chronic respiratory disease

District wise health system

| District | IMR (per 1000 birth) | Life expectancy at birth (Years) | CBR (per 1000 population) | TFR (per Women) | FAC (%) | Institutional deliveries (%) | | Complete immunisation (%) | Low birth weight (%) | Suicide rate (per lack population) | Number of beds (per lack of population) |
|---|---|---|---|---|---|---|---|---|---|---|---|
| | | | | | | Public | Private | | | | |
| Thiruvananthapuram | 11.0 | 75.2 | 16.4 | 1.6 | 71.8 | 67.3 | 32.6 | 81.6 | 11.4 | 33.4 | 238.0 |
| Kollam | 8.0 | 77.1 | 16.2 | 1.6 | 90.2 | 48.3 | 51.6 | 90.6 | 12.0 | 43.6 | 92.0 |
| Pathanamthitta | 8.0 | 76.7 | 14.5 | 1.5 | 84.8 | 27.5 | 72.4 | 91.4 | 18.0 | 32.9 | 97.0 |
| Alappuzha | 8.0 | 77.1 | 15.2 | 1.5 | 93.1 | 54.4 | 45.5 | 97.4 | 12.0 | 25.3 | 207.0 |
| Kottayam | 12.0 | 75.6 | 15.6 | 1.6 | 91.9 | 40.7 | 59.2 | 79.1 | 180 | 26.3 | 189.0 |
| Idukki | 20.0 | 72.4 | 17.0 | 1.6 | 82.1 | 25.4 | 74.5 | 90.8 | 15.0 | 51.7 | 96.0 |
| Ernakulam | 11.0 | 75.9 | 15.7 | 1.5 | 89.6 | 29.1 | 70.8 | 93.4 | 18.0 | 24.4 | 150.0 |
| Thrissur | 9.0 | 76.4 | 16.1 | 1.6 | 89.3 | 28.5 | 71.4 | 90.5 | 13.0 | 34.3 | 154.0 |
| Palakkad | 11.0 | 76.1 | 17.3 | 1.8 | 86.2 | 30.6 | 69.3 | 75.1 | 16.0 | 33.6 | 94.0 |
| Malappuram | 10.0 | 75.6 | 22.4 | 2.4 | 78.8 | 31.4 | 68.5 | 59.8 | 17.0 | 13.3 | 71.0 |
| Kozhikode | 12.0 | 75.4 | 17.4 | 1.7 | 93.1 | 54.3 | 45.6 | 90.9 | 17.0 | 23.3 | 210.0 |
| Wayanad | 22.0 | 73.5 | 19.5 | 2.0 | 90.4 | 54.2 | 45.7 | 82.3 | 30.0 | 46.7 | 122.0 |
| Kannur | 12.0 | 75.6 | 16.6 | 1.7 | 90.2 | 35.9 | 64.0 | 84.7 | 15.0 | 46.7 | 127.0 |
| Kasaragod | 10.0 | 75.7 | 18.9 | 1.9 | 75.4 | 11.1 | 88.8 | 87.4 | 15.0 | 24.0 | 77.0 |
| Coefficient of variation (%) | 36.0 | 1.7 | 12.0 | 14.4 | 7.9 | 39.7 | 24.9 | 11.2 | 28.3 | 33.6 | 39.9 |

*Source* Complied from Kerala Human Development Report, 2005

We also looked at health infrastructure.

This shows that indices are varying from district to district and an approach based on district statistics or metropolitan statistics can give a better approach than state-level statistics.

We felt that there are already systems in operation which have resulted in these achievements and our design of a smart living system for our study area shall take full advantage of that. There is a shortfall in certain health-related issues, and it will be solved by taking up this focus on smart living study. The second decision for the case study was we must look at holistically the health of the study area. This means we must study biophysical health, social health, psychological health and ecological health to design smart living around it with a system goal of holistic health for all.

We found two approaches when we deal with health. There is a biomedical approach to health where concentration is curative system while there are Autogenesis where we concentrate on healthy living and how to achieve that. Sigmund Freud concentrated on the sickness of mind while Abraham Maslow looked at creative and mentally health people who are self-actualising persons. There is also an environmentalist who concentrates on environmental issues to find a solution while there are ecologists who can design a well-balanced healthy ecosystem by design. In this study, we decided to follow salutogenic, Abraham Maslow and Ecosystem designers' approach. Another chapter of the book also presents how these approaches can be activated in space in different locations where it is needed or possibly based on what is discussed in this chapter on urban spaces and the public realm.

## 10 Design of Smart Living at Ekistics Scale 4–15

The design of smart living can be worked out based on the ecological and cultural system in a city. In a theocratic country probably, religion may be the focus of cultural and ecological system that affect life. For example, Judaism in Israel, Christianity in the Vatican, Buddhist way of life in Tibet in China and Islam in Saudi Arabia can be the respective focus of the cultural and ecological system of cities in Israel, Tibet in China, Vatican and Saudi Arabia respectively. In a secular country, there can be many foci and each federal state can follow that focus based on the statistical indicator of the relatively high performing aspect of life such as for example health sector which has thrived over few centuries in the Kerala state of India. Under secular constitutions like India, no one religion is there to shape the cultural and ecological system. Many religious ways of life coexist side by side in the same city. Many laws are framed for different religion for practice as per the constitution of the country. In such countries, the descriptive statistical analysis of achievement variables can be a guide.

The smart part explained with respect to ekistics scale 1–3 may become more complex for ekistics scale 4–15 but must be designed with respect to the primary focus of the cultural and ecological system. In secular countries like India living may be focused on some other aspect since in a secular country smart living is not focussed on majority religion, it is not politically correct since all religions are of equal status, irrespective of the relative size of the population. In a cultural system, all manifestations and patterns of living are equally valuable and no ranking. In this case, we must shift the primary focus to something else based on the local empirical reality. Again, this primary focus may differ from regions to regions within an urban agglomeration or city in a secular country. This brings about a diverse and exotic smart living. Whatever the focus the domain study shall proceed with the design for smart living, the domain can be a religious practice, constitution, or certain aspect of the culture values most as demonstrated by statistics say health along with consequent longevity as shown above. This is presented in Chap. "Geospatial Intelligence for Smart Living–Case of New Delhi" for Kozhikode Metropolitan Region.

Another approach may be the regenerative approach based on the needs of focus on the cultural and ecological system if these systems do not meet the expectation of the primary focus. Chap. "Role of Smart Land Scape Architecture in Smart Development of the UAE Case Study—Dubai" in this book present the research on these topics and this study start with the analysis of the domain to reach the design. This is presented in Chap. "Geospatial Intelligence for Smart Living–Case of New Delhi" again from Kozhikode Metropolitan Region.

## *10.1 Nature of Smart Living in Smart Cities*

Smart city is a knowledge-based city that develops extraordinary capabilities to be self-aware, how it functions 24 h and 7 days a week and communicate, selectively, in real time knowledge to citizen end users for a satisfactory way of life with easy public delivery of services, comfortable mobility, conserve energy, environment and other natural resources, and create energetic face to face communities and a vibrant urban economy even at a time there is National economic downturns.

Smart Living in Smart Cities is that gives people the opportunity to benefit from new ways of living. It involves original and innovative solutions aimed at making life more efficient, more controllable, economical, productive, integrated and above all sustainable.

"Smart Living" enhances the overall living experience of people across age cohorts and demographics. The Smart Living focus on improving an individual's ability to interact with electronic services, and on improving general wellbeing. The aim is to create a safer, more secure, accessible, and happy society.

What follows is a discussion approach based on Integrated health the prime focus of smart living in once chapter to demonstrate how smart living can be designed in cities.

## *10.2 Smart Living Focussed on Integrated Health*

This section explores the design approach to the smart living system in smart cities using the concept of Integrated health which considers the health of the cultural and ecological system of the city. Chapter "Geospatial Intelligence for Smart Living–Case of New Delhi" in this book expand this methodological approach to a study area Kozhikode Metropolitan Area of 2 million plus populations based on analysis, diagnosis and design of empirical data. The author does not advocate integrated health as the only way to design smart living for smart cities. There are more such foci or entry point for study available based on what foci impact more on the living system of a smart city. This will be evident in analysing the achievements of related variables on a time series.

### 10.2.1 Biophysical Well Being

The World Health Organization (WHO) defined health in its broader sense in its 1948 constitution as "a state of complete physical, mental, and social well-being and not merely the absence of disease or infirmity. "Chapters "Geospatial Intelligence for Smart Living–Case of New Delhi" and "Role of Smart Land Scape Architecture in Smart Development of the UAE Case Study—Dubai" expands this to include the ecological health of the study area since humans are part of the ecology, and ecological health, directly and indirectly, affects the wellbeing. Human wellbeing refers to the degree to which an individual, family, or larger social grouping (e.g. firm, community) can be characterized as being healthy (sound and functional), happy, and prosperous. The determinants of an individual's wellbeing can include characteristics of family, community, nation, and so forth.

Like human wellbeing, there is also a concept of ecological wellbeing which will be discussed later. Ecological health is very much related to human health, one influence the other. Physical well-being consists of the ability to perform physical activities and carry out social roles that are not hindered by physical limitations and experiences of bodily pain, and biological health indicators. Efficient delivery of biophysical health services can be designed using ICT and IOT system that make it smart delivery of health services on 24 h 7 days basis based on the accumulation of knowledge based on wearables, the analysis in the cloud instantly and instant actions triggered because of analysis.

### 10.2.2 Mental Wellbeing

Mental health is defined as a state of well-being in which every individual realizes his or her own potential, can cope with the normal stresses of life, can work productively and fruitfully, and is able to contribute to her or his community. Mental health includes emotional, psychological, and social well-being. It affects how we think, feel, and act. It also helps determine how we handle stress. Mandukya Upanishad and Gaudapada Karika with Asparsa Yoga discuss an analytical framework to work out mental wellbeing along with the Buddhist and other approaches discussed there. Generally, mental health is known based on mental diseases psychiatrists and psychologist diagnose it. There is a curative, preventive and promotive approach to mental health where the family, immediate primary groups around and community have a great role to play in addition to a psychiatrist and psycho analyst. The city can adjust itself for the mental needs of the inhabitant with better urban regeneration and urban restructuring involving land management. The role of society and family in addition to the curative assistance of doctors can be enhanced if IOT, ICT and e-democratic system of local communities are designed to meet the local area needs.

### 10.2.3 Social Wellbeing

Social wellbeing is the extent to which you feel a sense of belonging and social inclusion; since a connected person is a supported person in society. Lifestyles, ways of living together, value systems, traditions and beliefs are all important to our social wellbeing and quality of life and hence the paramount importance of the cultural system. Gautama Buddha, Jesus Christ and others have carefully laid out the protocol for social wellbeing in the religious texts. Social health involves your ability to form satisfying interpersonal relationships with others. It also relates to your ability to adapt comfortably to different social situations and act appropriately in a variety of settings. Spouses, co-workers and acquaintances can all have healthy relationships with one another. Each of these relationships should include strong communication skills, empathy for others and a sense of accountability. In contrast, traits like being withdrawn, vindictive or selfish can have a negative impact on your social health. Overall, social stress can be one of the most significant threats to a healthy relationship. The stress of individuals should be managed through proven techniques such as regular physical activity, deep breathing and positive self-talk. Places designed for social health shall be designed and made available at accessible points for interpersonal interactions.

For social health;

- Humans shall effectively develop relationships and maintain good social health; individuals must be willing to: "Give of themselves".
- This could include sacrificing time, effort, energy or money for others which include all living.
- Have adequate levels of self-esteem.
- Being mentally and emotionally secure with oneself can help an individual maintain healthy relationships.
- Establish a sense of identity.
- Sacrificing personal characteristics often results in less satisfying relationships, while acting like your true self will strengthen social bonds.

- Utilizing appropriate places in a city setting for developing a relationship; designed formal or informal.

All relationships will have some level of emotional involvement, also known as intimacy. Determining how intimate a relationship will become; is critical to long-term social health. While acquaintances or co-workers may have very little intimacy, family members and spouses often have intimacy levels high enough to be considered love. Characteristics of a healthy relationship include:

- Trust—those involved have faith in each other and will do what is best for the relationship.
- Compassion—the physical and emotional well-being of others in the relationship is considered important.
- Respect—sacrifices made for the relationship such as time, effort and money are acknowledged and valued.
- Acceptance—changing individual characteristics and personality traits is not an expectation.
- Reciprocity—the give and take within the relationship is relatively equal.

### 10.2.4 Ecological Wellbeing

Ecosystem health of a city represents the desired endpoint of environmental management. A healthy ecosystem is one that is sustainable, it can maintain its structure (organization) and function (vigour) over time in the face of external stress (resilience). The study of these attributes leads to a comprehensive assessment of ecosystem health.

Ecological health is a term that has been used in relation to both human health and the condition of the environment. Measures of broad ecological health, like measures of the more specific principle of biodiversity, tend to be specific to a homogeneous ecoregion or even to an ecosystem. Measures that depend on biodiversity are valid indicators of ecological health as stability and productivity (good indicators of ecological health) are two ecological effects of biodiversity. Dependencies between species vary so much as to be difficult to express abstractly. However, there are a few universal symptoms of poor health or damage to ecological system integrity: given next.

- The build-up of waste material and the proliferation of simpler life forms (bacteria, insects) that thrive on it—but no consequent population growth in those species that normally prey on them;
- The loss of keystone species, often a top predator, causing smaller carnivores to proliferate, very often overstressing herbivore populations;
- A higher rate of species mortality due to disease rather than predation, climate, or food scarcity;
- The migration of whole species into or out of a region, contrary to established or historical patterns;

- The proliferation of a bioinvader or even a monoculture where previously a more biodiverse species range existed.

## 11 Living and Smart Living Perspective (of Smart Home Ekistics Scale 1–3 Population 1–5)

We have discussed two ways of life prescribed by two religions, Advaita of later Hinduism and early Buddhism, but there are many Schools of Hinduism and Buddhism, having different philosophies and ways of living. You can design a smart living around relevant them. There is also an adaptation of these same religions in different countries advocating a modified way of living. For example, Tibetan Buddhism is different from Buddhism practised in Sri Lanka, and Thailand. There are also other religions like Islam, Christianity, Judaism, Jainism and so on with many subsects having a differing view of living. Daily living is influenced by these religions to a certain extent in the cultural and ecological system of the city. It is to a certain extent because of modifications required by the ecological as well as a cultural system which varies from city to city. The religious influence is there for the first in cultural system and practices also influence how people deal with the ecosystem. In a predominantly Islamic country like Saudi Arabia, Sunni Islamic sharia code is the Governing code, Buddhist country like Tibet, Sri Lanka and Thailand Governing code based on legislation is influenced by Buddhism and philosophy that shapes living in differing schools, in Vatican Rome Governing code is influenced by Christianity, India the Hinduism with special governance system for other religions and in Israel it is Judaism. We also discussed how Doxiadis, Geddes and Lefebvre view differently about the social, cultural spatial and ecological system that we call the system of living in a city. There are many more thinkers whose ideas can be added here, but not attempted.

None of the authors discussed above mentioned in their literature anywhere the concept of "smart living". However, there is a growing literature on smart living (in room/home)starting from the 1980s. They concentrate more on first three population units discussed by Doxiadis of human settlement scale (Table 1) discussed in earlier part of this chapter namely man with say smart wearables, smartphones, smart money transfer apps in his smartphone; room/home with smart devices such as intelligent WIFI operated and controlled switch, plugs, lights, door locks, fans etc. and ending up in smart room/home well managed by smart hub like Samsung smart thing smart hub, Wink Connected Home Hubor voice-operated devices like Amazon echo plus, Google Home Hub, Amazon show or similar or automated homes using artificial intelligence that uses programmable smart hub. With open source SDK of Amazon Echo and Samsung Smart Thing Smart Hub, any student of local technical institutions can design cheaper hubs for smart living projects in the study area and prepare many smartphone apps to be used by children, housewife and all. Using the hub, smart WIFI devices, can work in the desired way either by voice or vision control and/or controlled by cloud computing and artificial intelligence based on data mining activities in the cloud. The cloud server need not depend on Amazon cloud but can be in designated local universities. The inhabitant can programme these devices to desired scenes with a specific ambience the household members wants based on ever changing external stimuli sensed by IOTs. He can make many devices grouped and work together performing the desired objective simultaneously and synchronised. For example, without direct human intervention from time to time the microclimatic ambience of room and light parameters of the room can change to desirable and acceptable ambience based on activities within the room with external ever-changing microclimate or lighting condition. For example, Google Hub automatically adjust its display to suit the ambient light condition of the room automatically using light sensors so that it is comfortable during sleep and varying light conditions. There is not much effort on the smart living of other 12 population units discussed by Doxiadis and some of which we concentrate in this book, for example, neighbourhood, town, metropolis, mega city and meta city which is the subject matter of the book. No wonder the complexity of external environment adds many more variables to a high level of complexity to deal with to make it smart. The automatic control discusses above become very challenging when applied to other 12 population scales of Doxiadis calling for technological and social innovations continuously evolving.

The term "smart home" was first used in the 1980s by the American Association of Housebuilders, although there were hobbyist's pioneers constructing in the past the prototype of many varieties of smart, or wired homes much earlier largely not recorded [60]. A smart home is smart because it contains interactive technologies that make us able to control and manage in automata the home in response to its residents' immediate needs. This interaction is created by the interaction of a smartphone with the smart device using a Wi-Fi network using apps connected to the cloud. These needs can vary from time to time and season to season and can be single needs or multiple needs or simple needs or complex interrelated needs at some part of the home. Complexity increases when we add assistive technology if the inhabitant of the house is a senior citizen, or citizen requiring palliative care or childcare. Complexity changes if we want the house to be secured against terrorist attack, thieves and natural calamities.

The smart home development started already in the middle of the twentieth century, and by the end of the century, smart homes continue to evolve with newer technologies. By the beginning of the twenty-first century, smart homes are developed across the planet, particularly in the western countries with an overall goal of improving health and Quality of Living of all, especially vulnerable populations. UK, USA, Japan, Korea, China, Canada, Norway and Denmark are some of the leading countries that have developed and initiated such projects [61]. The technology in these projects is comprehensive and covers a wide range of solutions to different kind of users including elderly, frail elderly and physically disabled people and children addressing various needs as they develop. It is difficult to define smart home [62], one of the main reasons for this is that it is sometimes difficult to separate smart homes and smart home technology from assistive technology, technology for example for the old age and palliative care, emergency medicine or other similar systems and it seems that the terms are being used interchangeably. Therefore, a"smart home is a residence wired with technology features that monitor the wellbeing and activities of their residents to improve the overall quality of life, increase independence and prevent emergencies" [63]. A conceptual framework based on the different kinds of features and functions that smart homes could have existed. The framework consisted of five elements:

1. **Automation**: the ability to accommodate automatic devices or perform automatic functions;
2. **Multifunctionality**: the ability to perform various duties or generate various outcomes;
3. **Adaptability**: the ability to adjust (or be adjusted) to meet the needs of users;
4. **Interactivity**: the ability to interact with or allow for interaction among users; and
5. **Efficiency**: the ability to perform functions in a timesaving, cost saving and convenient manner.

This framework combined with the definition above, comprise the total framework of smarthomes and the technology. The essential idea of smart homes is that they offer the inhabitants a better quality of living, by offering autonomy and safety. There is a variety of technological solutions and opportunities that can be installed in homes and make them "smart". Smart home technology is defined as "the integration of home-based technology and services for a better quality of living" [64] and may include light control, temperature control, multimedia operation, window and doorway entry options and many more health-related solutions. It seems that there is only our imagination that is the limitation of the development of such technology. So, you can now design a smart house based on your culture of living. In their systematic review [65] propose that the different varieties of health-related smart home technology utilize a wide range of technology that are serving different goals and can be defined as six categories:

1. **Physiological monitoring**: Measurements of pulse, respiration, blood pressure.
2. **Functional monitoring/Emergency detection and response**: Measurement of activities, motion, food intake etc.
3. **Safety monitoring and assistance**: Automatic lighting, fall reduction/detection hazard detection.
4. **Security monitoring and assistance**: Intruder detection, security breach.
5. **Social interaction monitoring and assistance**: Phone call assistance, video communication, social connection.
6. **Cognitive and sensory assistance**: Medication reminders, lost key locator, reminder and management tools [63, p. 34].

Smart home raises a variety of questions related to ethical issues, accessibility and usability concerns.

### 11.1 Ethical Issues

Ethical issues are always present in the relationships between human beings and technology and not new. Automatic functions, video surveillance, and monitoring and other functions in smart homes are one of several functions that raise ethical concerns. It must be decided who has the legal right to view the content of the surveillance, monitoring or audiotapes and this content must also be secured to keep it from falling into the wrong hands. Amazon Echo is being marketed in India by Amazon. Anybody with Amazon Echo show from any part of the world can have a video call from anywhere any time. That person can see the room where the echo show is located which raises many ethical questions in their use. The same issue of privacy exists for medical statistics generated by IOTs. All things connected to the internet, or a local network, will leave an electronic trace. Protection of the network, telephone line and other communication systems are essential to maintaining the privacy and ethical concerns. An increase in the use of smart homes and smart home technology will have an impact on privacy and this is most important for recipients of IOT assisted care [61]. By installing such technologies, one must consider the effect it has on personal relationships (e.g. patient-physician relationships), moral and human ethical issues and other factors that may be involved. Appropriate legislation is another requirement for safeguard privacy.

### 11.2 Accessibility

Home Electrification is the primary requirement for smart living in smart houses and related devices that are generally attached to an electric system of the house. India's Prime Minister Narendra Modi declared in 2018 all houses are electrified in India with the last house being in Mizoram in the northeast of India. Now India can have smart homes for all its population provided technologies are cheaper to own. Although smart homes and smart home technology are considerably cheaper than they were a decade ago, it is still expensive for middle and low-income households. China has taken a lead in reducing the cost of such a device, but other countries keep it still expensive. This may be because of the required continuing technological development in developed countries and that technologies are constantly evolving with new advanced functions and innovative features which involves higher costs in developed countries with higher wage rates. Mass scale production base with an economy of scale is yet to come about in smart home technologies but it is possible in China and India with high potential consumer population. When installing such technologies, it may take an expert to do this which is also another constraint, but many are doing it yourself off the shelf. Servicing also requires experts and one must always consider this fact when the type of smart home technology installed. Advanced systems require software programming and future updates which are costly for many people. It is important that the government supports this, not only in large residential buildings in the municipality but also in private homes to those households who cannot afford to acquire such technology on their own. In Norway, the Housing Bank has been the Norwegian Governments most important institution for smart home delivery in the housing policy. Through grants, loans and guidance the Housing Bankhas helped with the development and construction of better housing solutions [66]. The Housing Bank is a good example of how technology and smart homes can be a reality for many people in India and China that cannot afford it themselves. Smart homes may save money for the government since it might be cheaper to provide smart home technology that allows elderly and sick to live longer at home rather put them in hospitals and nursing homes at Government expenditure.

## *11.3 Usability*

According to Lê et al. [63], another common concern is the limited familiarity that many elderlies and less educated have when it comes to advanced technology in smart homes. Unlike the new generations, today's elderly generation did not grow up with technology in their hands, something that may reduce their comfort and familiarity with such technologies. In a Korean quantitative survey conducted on 290 elderly participants [67] searched for the factors affecting elderly people's intention to use technology. The involved technology was video user-created content technology that facilitates social interaction through a video connection, like an advanced form of Skype. The study showed that it was central that technology had value and was practical and usable. There was a strong connection between the experience of utility, usability and enjoyment and it seemed that subjective assessments of their own physical health and resources had a direct impact on their intention to adopt the technology [68]. They conducted three focus group interviews with 15 elderly citizens in their study about older adults' attitudes towards and perceptions of smart home technology. The participants expressed concern about the user-friendliness of the electric devices installed. Another concern was related to the lack of human response inherent in these technologies and the need for a tailored training program for older people. To test the usability and acceptability of the technology in the European Union project SmartHome for All (SM4ALL), [69] conducted testing of the system with 31 elderly's participants aged between 47–91. The technology involved was an electrical brain activity(EEG) based brain-computer interface unit or BCI. The EEG-BCI unit measures and analyses the EEG in response to the participant controlling external devices. The results showed a good response and positive attitudes when it came to usability and acceptability, the elderly users seemed to be willing to give up some of their privacy to get better support at home. In addition to this, they also expressed that it would be an advantage if the technology was able to help them gaining control over tasks that previously had been difficult or impossible.

## 12 Design of Smart Living for Smart Cities

A synthesis of many considerations mainly strategic together leads to the design of smart living for smart cities. They are discussed below very briefly. They are enumerated below

1. Design for the development of an ecosystem for innovation in smart living
2. Spatial Design for Smart Living
3. Design of Smart Community
4. Design of Smart GIS for Smart Living
5. Design of GIS-based Spatial Decision Support System (SDSS)for E-Democracy and E-Governance for smart living
6. Design of GIS-based Zonal Plan Design including Hybrid Form-Based Codes for smart living
7. Design the ICT system for use in all smart living activities
8. Design of IOT system for all smart living activities
9. Design modifications to institutional arrangements for smart living implementation.

## 13 Strategic Actions for Smart Ecosystems of Innovation in Smart Living

Having discussed various aspects of the smart part of living, let us discuss the strategy for applying it to the ecosystem for innovations that leads towards a smart city is given below.

1. Invigorating the ecosystem for innovation to support the development, commercialization, scaling up and export of smart solutions. A smart solution to smart living can be both hardware and software in terms of smartphone apps as well as innovating smart protocol the community design. This means conceptualise, design, develop and deploy smart living devices for ecological and targeted to living by species, sex and age in collaboration with the smart community where mass scale adoption is visualised locally and the possibility of global demand.
2. Engaging the public and business and co-creating smart solutions with stakeholders to better meet the needs and fulfil the aspirations of the people: This means establishing mutually beneficial linkages with smart community and business and universities for field testing and adoption and opening business opportunities.
3. Cultivating a culture of embracing changes and collaboration for smart city development: this calls for continuing education as well as skills the smart community.
4. Integrating with city planning; This may involve changing the planning legislation or amending it. This also calls for educating the Planners on Smart Living.

5. Establishing a sound digital framework to support further improvement to the operations of urban infrastructures and provision of services; There shall be a program to augment backbone infrastructure of the existing ICT and IOT system in the annual budget of local bodies.
6. Providing support to reduce the digital divide and ensure inclusiveness of Smart City development. The policy of free provision of smart devices to the economically weaker households. Opening bank credit at concessional rate for smart device procurement by the smart community.
7. Equipping students and workforce with the knowledge and skills to seize opportunities in a Smart City: Skill development of students in emerging ecology based smart devices and its application. Creating an ecosystem for innovative start-ups.
8. Developing an innovative strategy for procuring smart solutions: Smart Community shall form a close alliance with local technological institutions. This alliance shall lead to discuss problems and design and implement innovative solutions from these technical institutions.
9. Improving preparedness, prevention, and mitigation of adverse events. Smart Community shall have a dedicated group of members to achieve this task year-round.

## *13.1 Spatial Smart Living Strategy*

1. Spatial means regional which happen to be a programming region of the metropolitan committee. There are no programming regions for megacities and meta cities in India and many countries. For example, Kozhikode Metropolitan Region is the constitutional body eligible for consolidated funds of union and State Government of Kerala and an elected body for its development. Although it is in the constitution it is not established yet in many states like Kerala.
2. There is a metropolitan committee to govern it as per the Town and Country Planning Act 2016 of Kerala (under revision). One of the goals of this committee shall be smart living for all and there shall be a budget to meet it.
3. It has constitutional functions to perform as per Act 2018 with definite membership. Membership may be expanded to include industry and academia which is not there.
4. The regional strategy of smart living means settlement, linkages and infrastructure strategy. Infrastructure includes ICT and IOT infrastructure for smart living.
5. Regional means consideration of all natural (ecological) and human resources and working out the resource-conserving strategy for smart living.
6. Regional means it is a polarized region with emerging settlement hierarchy and functional specialization which provides a spatial matrix to design smart living Strategy to increase metropolitan wellbeing and clearly point towards the sub-regional specific scope of work.

7. Spatial smart living strategy to avoid sub-regional metropolitan area disparity and make metropolitan wellbeing inclusive of poor and rich and middle class.
8. Strategy to increase metropolitan employment opportunity as related to smart living activities for locals and immigrants of all categories. Smart living is a generator of employment.
9. Strategy for use of natural and human resources and including strategy for import of these resources if not available locally.
10. Strategy for development of related institutions for well being.
11. Smart Living with Smart Economy, smart people, smart energy, smart environment, smart mobility, and smart governance in smart city framework.
12. Smart wellbeing and health initiative by deploying and using required ICT, and IOT.
13. Design for empowering households and community for smart living.

## *13.2 Smart Community Strategy*

1. A smart community is identified as Residential Welfare Association (RWA). A neighbourhood as conceived by planer in not easily spatially discernible but RWA is a registered entity under society act involved with the welfare of the local community.
2. RWA is below Ward Committee area, Panchayat, Towns and Metropolis. Those spatial unit above RWA are generally constitutional body but RWA is not. The constitutional body must draw annual plans as per Governmental protocol and spend their share of tax money collected by State and Union Government, as well as local bodies as per the directive of Financial Commissions and, are to elect members by regular election in periodic intervals under the constitutional body the Election Commission. RWA is also accountable to themselves under society act and the election is required to be conducted as per their registration document.
3. RWA consists of about 100 households mostly known to each other. In a Kozhikode Metropolitan Region, there may be around 20,000 plus RWA with a total population above 2 million.
4. Metropolitan Area as per constitution is required to prepare hierarchy of plans such as perspective/vision 20 years, midterm 5 years and annual plan tied up to the annual budget provision of the Metropolitan Area. This plan shall focus on smart living and include a component implemented by RWAs. The methodology for the plan will be like what is presented in Chap. 7 for Kozhikode Metropolitan Area.
5. Converting the RWA community of 400 to 500 people in a walkable area even for children is to be executed in a systematic way. The action program shall be drawn by RWA itself.

6. RWA shall be subjected to continuous training in skilling to meet the local needs as dictated by metropolitan plans for smart living execution. Households in Kerala are fully computer literate and have been declared as such. This computer literacy makes them familiar to operate computers and the internet and send emails etc. This Skilling shall be expanded to Geographic Information System Data Generation of RWA area GIS by using public domain GIS software and use the Spatial Decision Support System for smart living as discussed in Chap. "Geospatial Intelligence for Smart Living–Case of New Delhi". They may be trained to update RWA GIS once in every 3 months with new features and attributes as related to smart living.
7. In addition to GIS, they shall be educated as end users of various IOT system for smart living and assistive technologies for smart living by relevant experts from the Metropolitan Region or outside.
8. For continuous training for skills in smart living planning, execution and management RWA shall establish relevant linkages with Universities, Colleges and Vocational Schools adjacent to RWA.
9. RWA shall be trained to conduct E-Democracy at RWA area.
10. RWA shall assist the Government to implement legislation and acts both from central Government and State Government. This is discussed in Chap. "Geospatial Intelligence for Smart Living–Case of New Delhi".
11. The spatial Decision Support system shall be designed over GIS as a collaborative exercise of RWA and relevant smart living issues as discussed in Chap. "Geospatial Intelligence for Smart Living–Case of New Delhi" with the linkage already established with the relevant institutions.
12. There shall be a set of E-Governance SDSS and E-Democracy SDSS as discussed in Chap. "Geospatial Intelligence for Smart Living–Case of New Delhi" which RWA use daily for smart living.
13. For Government support for smart living planning, execution and management RWA shall establish relevant linkages with State and Central Government Offices in the Metropolitan Area. This will ensure full Government support for smart community action.
14. RWA shall assess from Metropolitan Plan their role on a 24 h 7 days basis in implementing Smart living,
15. Smart Community shall be organised to implement these protocols as discussed under section smart community in Chap. "Geospatial Intelligence for Smart Living–Case of New Delhi".
16. RWA shall organise free hosting for students of technical universities in their selected house who come forward to develop free smartphone apps for e-Democracy and e-Governance for smart living.
17. RWA shall organise free hosting for students of technical universities in their selected house who come forward to help as free volunteers for geriatric care and palliative care. For e-Democracy and E-Governance for smart living.
18. RWA shall develop in collaboration with cell phone operators the backbone of ICT and IOT system in their areas.

19. RWA shall organise free hosting for students of technical universities in their selected house who come forward to design and develop a public realm for smart living.
20. RWA shall identify selected houses for senior citizen congregation and activities for smart living.
21. RWA shall identify sites for locating and siting IOTs for specific use in conjunction with smart living with the help of relevant institutions in the metropolitan areas.
22. RWA shall identify server location for smart living cloud computing in relevant institutions.

## *13.3 A Smart GIS for Smart Living*

The GIS, geographic and attribute database creation and its updating shall be the job of the smart community at RWA level as explained in the previous para. GIS will be used extensively in smart living routines. Many examples are given in Chap. "Geospatial Intelligence for Smart Living–Case of New Delhi", but one example is given here. Imagine a smart living situation is sudden heart attack reported in a household of RWA. Immediate action required is to transport the patient to the nearest hospital where emergency care is available. To transport we require a car from the volunteer household nearest to that of patient and GIS shall track this house and the nearest hospital facility. Communication shall be sent to all concerned such as nearest volunteered household for transport, hospital and family of concerned patient instantly. GIS become smart when these messages are received with zero cost of time to all concerned by messaging and SMS facility of devices like Amazon echo and Amazon echo spot/show for video call automatically without human intervention or similar devices designed by local institutions.

## *13.4 Spatial Decision Support System (SDSS) Using GIS for Smart Living*

SDSS is integrated computer systems that support decision makers in addressing semi-structured or unstructured spatial problems in an interactive and iterative way with functionality for handling spatial and nonspatial databases, analytical modelling capabilities, decision support utilities such as scenario analysis, and effective data and information presentation utilities. As SDSS are multifaceted technologies that are manifested in many varieties, it is useful to look at some of the common traits that characterize them. Here, Stakeholder roles, their thinking process and special and unique characteristic of SDSS determine the design of SDSS for E-Smart Living Governance and E-Smart Living Democracy and these are diagrammatically represented

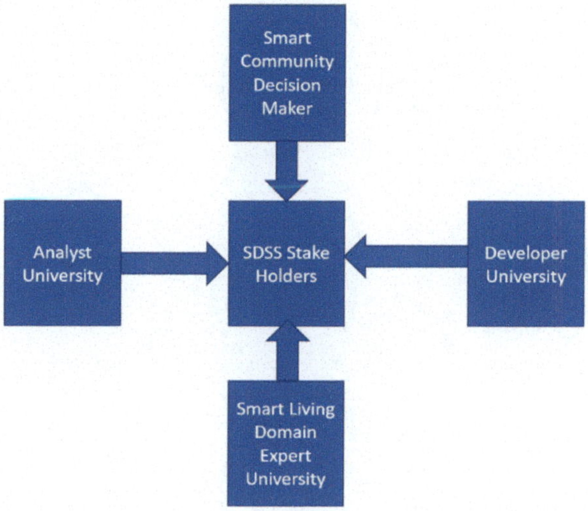


**Fig. 11** Stakeholders in SDSS. *Source* [75]

**Fig. 12** SDSS thinking process. *Source* Author

below. Smart community and local universities and colleges shall jointly identify SDSS and design smartphone apps for implementation of SDSS. More details of the examples are given in Chap. "Geospatial Intelligence for Smart Living–Case of New Delhi" (Figs. 11, 12, and 13).

- The above three diagrams shall guide the development of SDSS apps ported to smartphones.
- This will be applied to various legislations such as for example the Coastal Regulation Act etc. to produce smart living E-Governance Apps.
- This will be applied to various smart community activities related to smart living such as for example to produce smart living E-Democracy Apps.
- Design SDSS for E-Smart Living Democracy based on local issues where legislations are not enacted including development beyond regulations.
- These applications shall Identify the best use of land under acts like for example CRZ or best house or location for smart living activities.
- Design SDSS for E-Smart Living Governance is based on Central and State Level Governance.

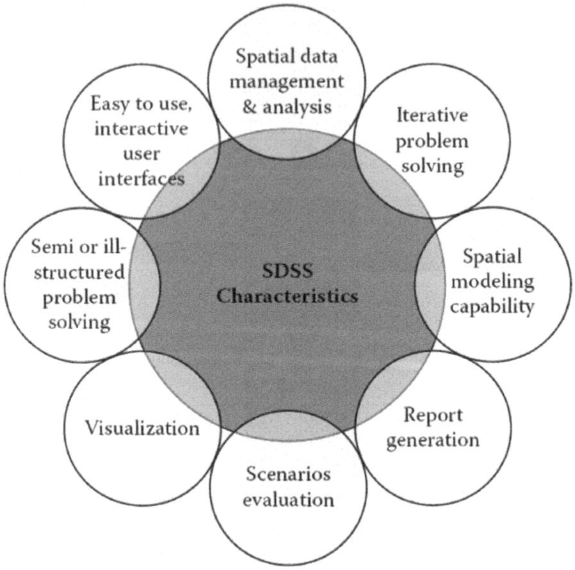


**Fig. 13** SDSS characteristics. *Source* [75]

## 13.5 Zonal Plan GIS for Smart Living

Zoning is a legal instrument of land use regulation used by local governments in most countries. The word is derived from the practice of designating permitted uses of land, based on mapped zones which separate one set of land uses from another. Theoretically, the primary purpose of zoning is to segregate uses that are thought to be incompatible. In practice, zoning is used to prevent new development from interfering with existing residents or businesses and to preserve the "character" of a community. Most zoning systems are flexible and have a procedure for granting variances (exceptions to the zoning rules), usually because of some perceived hardship caused by the nature of the property in question. Basically, urban zones fall into one of five major categories: residential, mixed residential, commercial, commercial, industrial and special (e.g. power plants, sports complexes, airports, shopping malls etc.).

Zoning codes have evolved over the years as urban planning theory has changed, legal constraints have fluctuated, and political priorities have shifted. Zoning Codes can be mapped giving clear direction to architects, urban designers, engineers and owner to avoid litigations and violations. Zoning codes can be expressed in alphanumeric characters. Zoning can be vertical or horizontal, homogeneous or heterogeneous such as mixed land use.

The various approaches to zoning can be divided into four broad categories: Euclidean, Performance, Incentive, and Design-based in addition to "Building

Block". Zoning is characterized by the segregation of land uses into specified geographic districts and dimensional standards stipulating limitations on the magnitude of development activity that can take place on lots within each type of district. Uses within each district are usually heavily prescribed to exclude other types of uses (residential districts typically disallow commercial or industrial uses). Some "accessory" or "conditional" uses may be allowed to accommodate the needs of the primary uses. Dimensional standards apply to any structures built on lots within each zoning district, and typically take the form of setbacks, height limits, minimum lot sizes, lot coverage limits, and other limitations on the "building envelope".

Euclidean zoning is utilized by some municipalities because of its relative effectiveness, ease of implementation (one set of explicit, prescriptive rules), long-established legal precedent, and familiarity to planners and design professionals.

Euclidean II Zoning uses traditional Euclidean zoning classifications (industrial, commercial, multi-family, residential, etc.) but places them in a hierarchical order "nesting", one zoning class, within another similar.

For example, multi-family is not only permitted in "higher order" multi-family zoning districts but also permitted in high order commercial and industrial zoning districts as well.

Protection of land values is maintained by stratifying the zoning districts into levels according to their location in the urban society (neighbourhood, community, municipality, and region).

Also known as "effects-based planning" performance zoning uses performance-based or goal-oriented criteria to establish review parameters for proposed development projects in any area of a municipality. Performance zoning often utilizes a "points-based" system whereby a property developer can apply credits toward meeting established zoning goals through selecting from a 'menu' of compliance options (some examples include: mitigation of environmental impacts, providing public amenities, building affordable housing units, etc.).

The appeal of performance zoning lies in its high level of flexibility, rationality, transparency and accountability. Performance zoning can be extremely difficult to implement and can require a high level of discretionary activity on the part of the supervising authority incentive zoning is intended to provide a reward-based system to encourage development that meets established urban development goals. Typically, a base level of prescriptive limitations on development will be established and an extensive list of incentive criteria will be established for developers to adopt or not at their discretion. A reward scale connected to the incentive criteria provides an enticement for developers to incorporate the desired development criteria into their projects. Common examples include FAR (floor area ratio) bonuses for affordable housing provided on-site and height limit bonuses for the inclusion of public amenities on-site. Incentive zoning allows for a high degree of flexibility but can be complex to administer. The more a proposed development takes advantage of incentive criteria, the more closely it must be reviewed on a discretionary basis.

Form-based zoning regulates not the type of land use, but the form that that land use may take. For instance, form-based zoning in a dense area may insist on low setbacks, high density, and pedestrian accessibility among other things. Form-based zoning also may specify desirable design features, however, when form-based codes do not contain appropriate illustrations and diagrams, they have been criticized as being difficult to interpret as another example, in a largely suburban single-family residential area, uses such as offices, retail, or even light industrial could be permitted so long as they conformed (setback, building size, lot coverage, height, and other factors) with other existing development in the area.

Hybrid Form-Based Codes is inclusive of zonal codes. There are few Form-Based Code Practitioner who considers this as appropriate. Others consider what is important is form and not functions. Functions can vary. The author recommends that smart components of spatial zoning shall also be part of the zoning for smart living.

## 14 Major Questions of Smart Living in Cities, Metropolises, Megacities and Meta Cities

1. How to realize existing institutional capacity to implement Smart Living for Smart Cities program in Metropolitan Areas and design ways and means to plan and implement. I do not recommend special purpose vehicles which may weaken existing constitutional bodies.
2. How to use the Constitutional power and responsibility vested in the Metropolitan Planning Committee to plan, design, implement and govern for a smart living for smart cities.
3. What are the existing provisions in terms of infrastructure for smart living in study areas, gaps, potential and SWOT?
4. What constitutes place making for a design for need-based Physical, Mental, Social and Ecological Health for smart living in the Metropolitan area? (This is because we assumed the primary focus health for our Kozhikode study given in Chap. 7).
5. What constitutes place making for a design for cultural needs for smart living in the Metropolitan area?
6. What constitutes place making for a design for a sharing community production and services needs for smart living in the Metropolitan area?
7. How to prepare and use various types of zonal codes designed largely by the community including hybrid Form-Based Codes and design modules to implement smart living strategies. Chapter 8 will answer it.
8. How to design and use ICT and IOT system for a smart living (Vide Chap. "Geospatial Intelligence for Smart Living–Case of New Delhi").
9. How to design E-Democracy for smart living including a spatial decision support system (Vide Chap. "Geospatial Intelligence for Smart Living–Case of New Delhi").
10. How to design E-Governance in Smart Living including spatial decision support system (Vide Chap. "Geospatial Intelligence for Smart Living–Case of New Delhi").

## 14.1 The Basic Smart City Concept and Its Implementation of Smart Living in Cities and Metropolises as Against Smart Home

There is a great difference between the concept of smart in a home and metropolis, mega city and meta-city. The difference is like the difference between the discipline Architecture, Urban Design, Urban Planning and Regional Planning. When we move to a higher level of population scale of the settlement, an individual cannot design the smart cities himself as he did in a smart home as he moves from smart home design to smart metropolis design for smart living. While the smart city was defined as a six-component system as given in Fig. 1, Smart living also become a six-component system corresponding to smart city system for living and a total design means designing the six components as given in Fig. 7.

1. Smart Living Communities: They are the community entrusted with the main task of executing the enabling act of smart living based on the domain selected and in Chap. "Geospatial Intelligence for Smart Living–Case of New Delhi" example it is Integrated health. In our example, it is Residential Welfare Association. Their effective role using IOT, ICT, E-Democracy and E-Governance as related to domain example smart and integrated health requires constant updating of required skill sets related to domain health and skills for assistive and enabling technological advancements such as IOTs and ICTs. Relevant institutional linkages such as with Universities in the study area provide for it. The smallest existing institution like Residential Welfare Association with about 100 household averages is a good selection of the area of the smart community. They need to be made a legal entity by registering under the relevant actors in the case in the India Society Registration Act. In the act, a clear articulation of all their tasks as related to enabling smart living shall be stated based on studies (Fig. 14).
2. Smart living related to Mobility: Smart mobility uses ICT and IOT and modes of mobilities are organised as per tasks enumerated in the Terms of References of the Smart Community for smart living. There can be emergency mobility and normal mobility. Normal mobility can even be slower walking and cycling if one must enjoy nature. Smart community design such mobility system and implement it as ICT enabled mobility using a smartphone.
3. Smart Living Economy: There is an economy as related to smart living related to industries, commerce, transportation, recreations, tourism and religious festivals. The tasks of the smart community are to convert this economy to smart using ICT and IOT and reap the benefit.
4. Smart Living Governance: There are several legislations both at the Union Government and State/Federal Government for activities related to smart living. The task of the Smart Community is to convert it to E-Governance system operated by smartphones with the help of linking institutions and their infrastructure. The smart community assumes an assistive role through e-smart living governance to the Government without encroaching the role the constitution has given to government institutions. Then smart community becomes an enabler of E-Smart

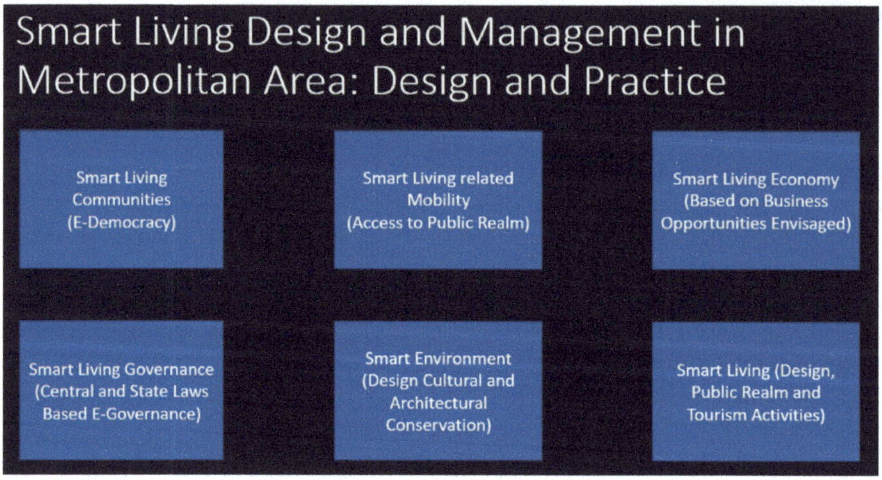


**Fig. 14** Six components of the smart living system

Living Governance. This may involve designing a GIS-based Spatial Decision Support System as related to acts.

5. Smart Environment: The task here is how do we convert environment management as related to ecological, cultural and architectural conservation related smart living activities to smart using ICT and IOT designs by the smart community. This also requires knowledge-based assistance by local competent institutions. Micro-level data collection on everchanging environment parameters like for example water quality using IOT and cloud computing shall be the responsibility of the smart community.
6. Smart Living; Here living as dictated by domain focus is converted to Smart using ICT, IOT, E-Democracy and E-Governance. Spatial design and ICT and IOT implementations in all public realm for tourism and other living activities shall be attempted. The local institutions will be solicited by the smart community to implement this component by activating the smart realm.

What is discussed above is diagrammatically presented as methodology below? (Fig. 15).

It can be seen here that the Smart Home approach is very different from the smart living approach for large size settlements. The scale of complexity increases and many institutions must work together using ICT and IOT.

## *14.2 Implementation*

An action plan for a smart community shall be there to start with. In the case of a metropolitan area with say 2 million plus population like Kozhikode Metropolitan Region, there shall be 20,000 RWAs which can be converted to the smart community

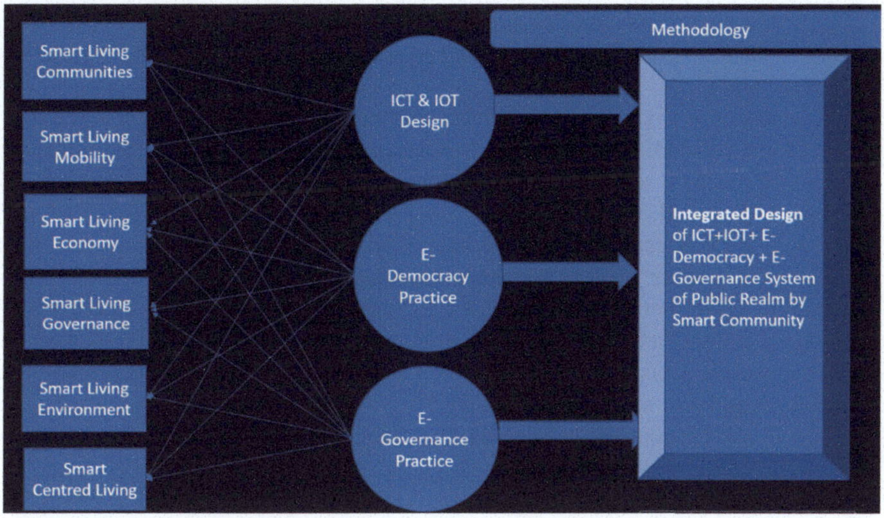


**Fig. 15** Integrated methodological design for smart living

with continuing upgradation of their skill set and knowledge base oriented towards smart community actions for smart living. This is possible only if local institutions take lead in this program. RWA should have a smart living plan of their own jointly formulated by local institutions and 100 plus household members of the RWA for the smart community based on local realities and issues. It can focus on a theme such as health. This can be based on a Metropolitan Regional Plan for a smart living or for a neighbourhood or even for a community. Metropolitan planning Committee shall take smart living as one of its goals.

Planning for Smart Living in a Smart City essentially is answering community how the following tasks may be performed in the plan for smart living. Chapter "Geospatial Intelligence for Smart Living–Case of New Delhi" shows one example of executing these tasks.

1. What is the concrete term are the Concepts of Smart Cities, Smart Living, Smart Communities, Smart Living E-Governance, Smart Living E-Democracy for the smart community for the study area?
2. What are the examples of smart living which we can study as replicable good examples of smart living in smart cities programs from India and abroad?
3. What should be the Aims of regional or community study achievable, what are its objectives, what methodology to use, Service Area Delineation, Area Profile with respect to the smart living, Mapping and secondary data collection.
4. How to assess existing Living Situation Physical, Mental and Social Health, Existing land use standards as related to living, Public Realm for Physical, Social and Mental Health of all age group, Existing Public Health Infrastructure, Existing Access and Mobility to the public realm and health infrastructure, Existing Environmental Health.

5. How to assess and specify problems and issues of living, Physical, Mental and Social Health, Smart City Policies and standards, Health Policies and Standards, SWOT and Gaps.
6. Assessment of Living in Study Metro- from domain foci example health statistics, land use studies, environment studies, mobility studies of public realm and comparison with selected metros with the high quality of life.
7. What Geospatial analysis of study area using RS and GIS for Smart Living of ecology and ecosystem of the area, Geospatial analysis of Environmental Governance using Remote Sensing and GIS and legislative norms and direction spatial Decision criteria such as CRZ etc. can be used for the above objectives.
8. How to conduct a Study of smart living and related quality of life indicators.
9. What are the Existing Land use needs and provide for human needs by age groups and income for a smart living?
10. How to the assessment of ecological needs and provide for human needs for smart living.
11. What are the Institutional set up for Urban Planning, Urban Development and Urban Management in the study area, Constitutional responsibility of the Metropolitan Planning Committee, Municipal Corporation, Municipality and Urban Panchayat to deliver smart living? Financial Capability for urban development for smart living, Capacities and gaps in executing Smart Living Projects in the study area, SWOT analysis of Institutions for smart living projects and suggested remedies.
12. How to briefly Summarise Assessment of Issues and Gaps.
13. How to conduct a SWOT analysis and Vision of Smart Living.
14. How to design Land suitability analysis using RS and GIS for the ecological needs of the city.
15. How to conduct Identification of location for public realm for local needs.
16. How to assess the Economic Opportunities in the Public Realm, Existing Public Realm Governance.
17. How to decide on Design standards and modules for the smart living of the public realm.
18. How to conduct Designation of Euclidian, Performance and Incentive and Form-Based Codes for Smart Living.
19. How to design E-Governance for Smart Living based on existing legislation as related to living.
20. How to design E Smart Living Democracy, Principles, Concept, Methodology, Case studies, Proposals for Smart Living for executing smart living in the area.
21. How to design IOT and ICT System for Smart Living using existing infrastructure.
22. How to frame Proposals: Urban Land Use Standard for Smart Living, Zoning standards for smart living, Hybrid Form-Based Codes and Smart Living Standards.
23. How to frame Proposals: Smart Living Design Modules for Hierarchy of settlements.

24. How to frame Proposals based on Multicriteria Spatial Decision Support System for Smart Living Spaces.
25. How to frame Proposal: Identification of living spaces and creation by urban land management for humans by age group and design criteria, Identification of living spaces for other living by type and design criteria.
26. How to frame ICT and IOT for Smart Living, IOT for smart living skills, IOT groups for Smart Living skills, IOT scenes for Smart Living skills.
27. How to frame ICT proposals and its use for smart living, Infrastructure proposals for ICT and IOT.
28. How to create Smart Living Communities.
29. How to design Smart Community for smart living management, Legal structure, Financing Strategies, Bylaws.
30. How to create Smart Community design by smart living activities and registration as a society with specific site-specific objectives of smart living.

## 15 Smart Living Perspective of the International Collaborative Research for the Book "Smart Living for Smart Cities"

This book based on international collaborative research; is aimed at developing state of the art of "Smart Living"; one of six components for Smart Cities. The others being smart people, smart economy, smart environment, smart mobility and smart Governance. The smart living in the smart city can only be designed and executed by an active role of Smart People, local institutions of higher learning and Government and is a joint and synchronous effort of E-Democracy, E-Governance and ICT-IOT system in a 24 h 7-day framework on all living activities domains in any smart cities. The smart living is a practice that uses information and communication technologies, Internet of Things, Internet of Governance (E-Governance) and Internet of People (E-Democracy) along with conventional domain-specific tools of many aspects of living by age cohorts to realise the coordinated, effective and efficient management, development, and conservation that improves ecological, social, biophysical, psychological and economic welfare in an equitable manner without compromising the sustainability of development ecosystems and stakeholders. This book will present many city communities, and neighbourhood case studies centred on domain-specific smart living components. It will also showcase tools for smart living.

There are several research studies presented in this book. The geographic coverage of city studies is as follows. There are some studies from India and some from outside India namely, two each from USA and Nigeria and one each from China, UK, Israel, and UAE. Some of the studies generally follow Doxiadis settlement scaling from One man to Ecumenopolis, but not all the 15 types of settlement are covered. The focus of smart home was from Anthropos to home and it is not covered in the book. There are studies from Communities and Neighbourhood, some areas within cities, the whole metropolises and then megacities and meta cities.

The city as a cultural system generally gravitates around a religion since religion is mostly a way of life or a dominant value system guiding the living. This is there with Judaism of Israel case study of Galilee region. Although India with 10 studies is having a higher number of population of Hindus, there are no Indian studies centred around Hindu way of life discussed in the earlier part of this chapter since India is by the constitution is a secular country with full liberty to practice any religion. So Indian cities are secular cities with different governing codes for different religions.

Following functional characteristics and interlinkages for living discussed by Geddes earlier, there are several studies focussing on smart living from one widow of function. I tabulate such studies below in Table 2 based on Thematic divisions of chapters (Table 6).

There are 18 studies in this book which includes multiple city studies by three study teams (from the USA, India, China) out of total 18 from 7 countries (10 from India, 2 each from USA and Nigeria, one each from UK, UAE, Israel and China). This is a probably a pioneering opportunity for these 18 study teams to work on smart living in metropolis, megacity and meta city population scales as well as in smart community and smart neighbourhood level in contrast to smart wearables fitted person with assistive technologies, smart room or home or Architecture.

## 16 Conclusion

Planning, managing and development of Smart living for smart cities is based on some important principles of smart cities.

1. Smart living involves sharing freely human resources for smart living. All living activities are fully supported and shared by all involved locally without monetising this sharing. It is the generator of strong community feeling locally and making people aware of all happenings around them and find effective roles to face it.
2. Smart Living shall be based on the demand of local culture, local ecosystem and dominant values of the community such as religion, health etc.
3. Smart Living strengthens and empowers the local community by use of ICT, IOT, E-Democracy and E-Governance designed as per local needs by local people and institutions.
4. Smart living is based on a holistic and comprehensive view of life.
5. The marginal cost of sharing is kept near zero to make smart living sustainable.
6. Smart Living is essential to work with the local smart community.
7. Smart living is possible only if all local institution of higher education helps the local community in their continuing education and skilling to manage smart living as well as local planning.
8. Smart living requires sharing facility and services of local institutions to help create a smart living.

**Table 6** Thematic division of studies in the book "Smart Living for SmartCities"

| Chapters | Thematic area | Focus | Country | Study City |
|---|---|---|---|---|
| 1 | Introduction | | | |
| 2 | Smart open spaces for smart living | City | China | Hong Kong |
| 3 | Just smart living | City | India | Delhi |
| 4 | Regulated smart living | City | India | Dehra Dun |
| 5 | Sustainable healthy living | City | India | Kozhikode |
| 6 | Smart spaces for smart living | City | India | Kozhikode |
| 7 | Sustainable community cultural spaces for smart living | Community | India | Smart Initiatives in Sustainable Historic Capital city of The Himalayas-SHIMLA |
| 8 | Inclusive smart living | Community | India | Surat |
| 9 | Crimeless smart living | Community | India | Vijayawada |
| 10 | Smart living connecting communities, religions and economy | Community | Israel | Galilee Region |
| 11 | Smart living in smart cities in the United States | Community | USA | Cincinnati, Cleveland, Pittsburgh, Atlanta, Tampa, etc. |
| 12 | Bhopal smart living | Tool | India | Bhopal |
| 13 | Geospatial intelligence for smart living | Tool | India | Surat |
| 14 | Spatial intelligence for Smart Living | Tool | India | Thiruvananthapuram |
| 15 | Smart neighbourhood for smart living | Community | USA | St Louis |
| 16 | Smart lagos living | Tools | Nigeria | Lagos |
| 17 | Smart lagos mobility for smart living | Tools | Nigeria | Lagos |
| 18 | The smart landscape for smart living | Tools | UAE | Dubai |
| 19 | Smart mapping for a smart living | Tools | UK | Derby |
| 20 | Conclusion | | | |

9. Smart community assists in Governance as related to smart living by using smartphone based on E-Governance Applications for a smart living using the help of local institutions of higher learning.
10. Smart Community assist activating smart living by use of use smartphone-based E-Democracy of Applications for a smart living using the help of local institutions of higher learning.

11. IOT system for smart living shall be developed and managed by the smart community with the help of local institutions of higher learning.
12. Server for smart living apps will be in an institution of higher learning and research locally.
13. Wherever possible data mining and artificial intelligence will be applied for the speedy delivery of smart living.

## References


1. Vinod Kumar TM and Associates (ed) (2014) Geographic information system for smart cities. Copal Publishing Group, New Delhi
2. Vinod Kumar TM (ed) (2015) E-Governance for smart cities. Springer, Singapore
3. Vinod Kumar TM (ed) (2016) Smart economy in smart cities. Springer, Singapore
4. Vinod Kumar TM (ed) (2017) E-Democracy for Smart Cities. Springer, Singapore
5. Vinod Kumar TM (ed) (2018) Smart metropolitan regional development: economic and spatial design strategies. Springer, Singapore
6. Vinod Kumar TM (ed) (2019) Smart environment for smart cities. Springer, Singapore
7. Doxiadis CA (1967) Health of mankind. In: Wolstenholme G (ed). Churchill, London, pp 178–193
8. Doxiadis CA (1968) Ekistics. pp 374–394
9. Doxiadis CA. Emergence and growth of an urban region
10. Doxiadis CA (in press) The future of human settlements, Speech delivered at the 14th Nobel Symposium, Stockholm, Sweden, September 1969. Wiley, New York
11. Doxiadis CA (1968) Ekistics: an introduction to the science of human settlements. Oxford University Press, London, pp 27–31
12. Doxiadis CA (1968) Science 162:326
13. Doxiadis CA (1966) Between dystopia and Utopia. Trinity College Press, Hartford
14. Hall ET (1959) The silent language. Doubleday, Garden City, NY; (1966) The hidden dimension. Doubleday, Garden City, NY
15. Christaller W (1993) Die zentralen Orte in Suddeutschland. Fischer, Jena
16. Geddes P (1949) Cities in evolution, vol 27. London, pp 109–123
17. Geddes SP (1915) Cities in evolution: an introduction to the town planning movement and to the study of civics. Williams & Norgate, London
18. Mumford L, Geddes P, Novak FG (1995) Lewis Mumford and Patrick Geddes: the correspondence. Psychology Press
19. Geddes P (1920) The life and work of Sir Jagadis C. An Indian, Bose
20. Geddes SP (1979) Civics: as applied sociology. Leicester University Press, Leicester, pp 75–97
21. Buser M (2012) The production of space in metropolitan regions: a Lefebvrian analysis of governance and spatial change. Planning Theor 11(3):279–298
22. Carp J (2009) "Ground-Truthing" representations of social space: using Lefebvre's conceptual triad. J Planning Educ Res 28(2):129–142
23. Allen J, Pryke M (1994) The production of service space. Environ Planning D Soc Space 12(4):453–475
24. Borden I, Kerr J, Rendell J, Pivaro A (eds) (2001) The unknown city: contesting architecture and social space. MIT Press, Cambridge
25. Harland Bartholomew and Associates (1928) A plan for the city of Vancouver. In: Harvey D (2012) Rebel cities: from the right to the city to the urban revolution. Verso Books, New York
26. Andres L (2013) Differential spaces, power hierarchy and collaborative planning: a critique of the role of temporary uses in shaping and making places. Urban Stud 50(4):759–775. Borden I, Kerr J, Rendell J, Pivaro A (eds) (2001) The unknown city: contesting architecture and social space. MIT Press, Cambridge

27. Abercrombie P (1933) Town and country planning. Oxford University Press, Oxford
28. Allmendinger P (2000) Planning in postmodern times. Routledge, London
29. Burnham D, Bennett EH (1909 [1993]) Plan of Chicago. Princeton Architectural Press (reprint)
30. Chadwick G (1971) A systems view of planning: towards a theory of the urban regional planning process. Elsevier, London
31. Feinstein S, Campbell S (eds) (2011) Readings in planning theory, 3rd edn. Willey-Blackwell, Oxford
32. Fincher R, Iveson K (2008) Planning and diversity in the city: redistribution, recognition and encounter. Palgrave Macmillan, Basingstoke
33. Forshaw JH, Abercrombie P (1943) County of London plan
34. Fyfe N (1996) Contested visions of a modern city: planning and poetry in post-war Glasgow. Environ Planning A 28(3):387–403 [Glasgow Corporation (1945) First Planning Report]
35. Groth J, Corijn E (2005) Reclaiming urbanity: indeterminate spaces, informal actors and urban agenda setting. Urban Stud 42(3):503–526
36. Bhattacharya V (1925) The Mandukya Upanishad and the Gaudapada Karika. Indian Hist Q I:119–125, 295–302
37. Subbiah AV (1933) The Mandukyopanisad and Gaudapada. Indian Antiquary 62:181–193
38. http://tripurashakti.com/mandukya-upanishad-12-verses-on-aum/
39. Chennakesavan S (1960) The concept of mind in indian philosophy. Asia Publishing House, Bombay. Second Edition 1980 (MLBD) Delhi
40. Bhattacharya V (1940) "Gaudapada". Indian Hist Q XIV:392–397 (Also, in the Winternitz Memorial Volume, Lahore 1940)
41. Bhattacharya V (1940) "Asparsa-Yoga". In: Mehar Chand LachhmanDas, Woolner commemorative volume, Lahore, pp 19–20
42. Chakraborty NB (1957) Gaudapada: his work and views. Calcutta Rev 145:89–96
43. Conio Caterina (1971) The philosophy of the Mandukya Karika. Bharatiya Vidya Prakasan, Varanasi
44. Divanji PC (1940) Gaudapada's Asparsayoga and Samkara's Jnanavada. Poona Orientalist 4:149–158
45. Majumdar JL (1947–8) "Gaudapada's Karika". J Ganganatha Jha Res Inst (Allahabad) 5:203–227, 347–379; 6 (1948–9):65–87
46. Majumdar JL (1947) Philosophy of Gaudapada. Indian Hist Q 23:1–16
47. Majumdar JL (1950–1) "Philosophy of Gaudapada (alata-santi-pra- karanam). J Ganganatha Jha Res Inst 8:115–129, 233–249, 335–371; 9 (1951–2):11–25
48. Sarma BNK (2000) Further light on Gaudapada-Karikas. Rev Philos Religion (Poona) 3(1):45–55
49. Sarma BNK (1931) New light on the Gaudapada-Karikas. Rev Philos Religion (Poona) 2(1):35–56
50. Sarma BNK (2000) Still further light on the Gaudapada-Karikas. Rev Philos Religion (Poona) 4(2):174–195
51. Sarma BNK (1936–7) Upanishadic theory of Gaudapada's Karikas. Poona Orientalist 1(2):27–38
52. Subbiah AV (1935) Gaudapada's Agamalastra. Indian Hist Q 11:783–790
53. Harvey P (1990) An introduction to buddhism. University Press, Cambridge
54. Conze E (1951) Buddhism: its essence and development. Cassirer, Oxford
55. Gethin R (1998) The foundations of buddhism. University Press, Oxford
56. Bluck R (2002) The path of the householder: buddhist lay disciples in the Pāli Canon. Buddhist Stud Rev 19(1):1–18
57. Fronsdal G (2005) The Dhammapada: a new translation of the buddhist classic with annotations. Shambala Publications, Boston
58. The Pāli Canon and commentary (in Pāli) from the Dīgha, Majjhima, Saṃyutta, Aṅguttara, and Khuddhaka Nikāyas—PTS hardcopy editions, and online editions from http://www.tipitaka.org/romn/ (based on the CSCD, published by Vipassana Research Institute), http://www.tipitakastudies.net/ (also based on the CSCD, published by the Dhamma Society), and http://www.bodhgayanews.net/pali.htm (based on the Buddha Jayanti Tripitaka, published by the Journal of Buddhist Ethics)

59. Gombrich R (1988) Theravada buddhism: a social history from ancient Benares to modern Colombo. Routledge, London. https://doi.org/10.4324/9780203310878
60. Harper R (2003). Inside the smart home: ideas, possibilities and methods. In: Harper R (ed) Inside the smart home, vol 1. Springer, London, pp 1–7
61. Teknologirådet (2009) Fremtidens alderdom on any Teknologi (Rapport: 2009). Teknologirådet, Oslo. ISBN 978-82-92447-29-1
62. Chan M, Campo E, Estève D, Fourniols HY (2009) Smart homes—current features and future perspectives. Maturitas 64:90–97. https://doi.org/10.1016/j.maturitas.2009.07.014
63. Lê Q, Nguyen HB, Barnett T (2012) Smart homes for older people: positive aging in a digital world. Future Internet 4:607–617. https://doi.org/10.3390/fi4020607
64. Van Berlo A (2002) Smart home technology: have older people paved the way? Gerontechnology 2(1):88–87
65. Demiris G, Hensel BK (2008) Technologies for an ageing society: a systematic review of "smart home" applications. In: Yearbook of medical informatics 2008: access to Health Information, vol 1, pp 33–40
66. SINTEF (2013) Teknologistøtte I egenbolig. Available from: http://www.sintef.no/Presserom/Forskningsaktuelt/Teknologistotte-i-egen-bolig/
67. Ryu MH, Kim S, Lee E (2009) Understanding the factors affecting online elderly user's participation in video UCC services. Comput Human Behav 25(3):619–632. https://doi.org/10.1016/j.chb.2008.08.013
68. Demiris G, Rantz MJ, Aud MA, Marek KD, Tyrer HW, Skubic M, Hussam A (2004) Older adults' attitudes towards and perceptions of smart home technologies: a pilot study. Med Inf Internet Med 29(2):87–94. https://doi.org/10.1080/14639230410001684387
69. Aeillo M, Aloise F, Baldoni R, Cincotti F, Lazovik A, Mecella M, Pucci P, Rinsma J, Santucci G, Taglieri M (2011) Smart homes to improve the quality of life for all. In: 33rd annual international conference of the IEEE EMBS, Boston, Massachusetts USA, pp 1777–1780

# Community Studies

# Urban Transformation for Sustainable Growth and Smart Living: The Case of the Atlanta Beltline

**Sweta Byahut, Sudeshna Ghosh, and Calvin Masilela**

**Abstract** The Beltline project in Atlanta combines elements of urban greenway planning, parks and recreation, pedestrian and bike friendly trails, public transit, and other facilities connecting over 40 neighborhoods. This is bringing about transformative urban regeneration in those neighborhoods, with large-scale economic investments and growth in previously neglected areas and brownfields. The Beltline has generated considerable interest among urban scholars. However, most studies have focused on specific aspects of the Beltline development in detail. This study uses extensive literature review and analyzes census data while taking a comprehensive look at the following smart living perspectives: urban health and wellbeing, smart public transportation, economic development and brownfields, and managing affordable housing during neighborhood change. While the Beltline has brought about considerable growth and investment, and created city level green spaces and trails, the process has been accompanied by gentrification and displacement that has disproportionately affected the racial minorities and poor neighborhoods.

**Keywords** Atlanta beltline · Urban regeneration · Gentrification · Affordable housing · Trails and greenways

## 1 Introduction

In recent years there has been a renewed focus and increasing popularity of urban greening projects. These are envisioned to fulfil multiple objectives from transforming dilapidated neighborhoods, making inner city neighborhoods more attractive,

S. Byahut (✉)
Auburn University, Auburn, AL, USA
e-mail: swetabyahut@auburn.edu

S. Ghosh · C. Masilela
Indiana University of Pennsylvania, Indiana, PA, USA
e-mail: ghoshsn@iup.edu

C. Masilela
e-mail: cmasilel@iup.edu

 
T. M. Vinod Kumar (ed.), *Smart Living for Smart Cities*,
Advances in 21st Century Human Settlements,
https://doi.org/10.1007/978-981-15-4603-7_2

bringing in investment in inner cities, encouraging economic development, promoting healthy living for residents by improving access to parks, green spaces and trails, encouraging more mixed-use development, increasing the value of inner city lands, and for promoting sustainable urban development. Some of these successful projects include The High Line in New York, the Promenade Plantée at Paris, Reading Viaduct at Philadelphia, the Bloomingdale Trail in Chicago, Los Angeles River Revitalization, The Midtown Greenway at Minneapolis, and the Madrid Rio in Spain.

Rarely has an urban planning project in the United States (U.S.) generated as much interest in recent times as the Atlanta Beltline; having a far-reaching impact on numerous quality-of-life aspects and providing extraordinary opportunities. The Atlanta Beltline is one of the largest urban greenway and redevelopment projects currently under implementation in the Atlanta metropolitan region in the State of Georgia (Fig. 1). The project visualizes transformation of 45 neighborhoods along an underutilized urban railroad stretching over 22-miles, developing a range of parks, trails, transit facilities, and attracting large-scale commercial, residential and mixed-use developments (Fig. 2). This ambitious project is estimated to have a total cost of $4.8 Billion and will be funded through a variety of public and private sources, including the Atlanta Beltline Tax Allocation District (TAD), the City of Atlanta, private investments, philanthropic contributions, county, regional, state and federal grants, and public private partnerships (www.Beltline.org). The overall purpose is to improve the quality-of-life of urban residents, improve public health outcomes by increasing walkability, create parks and greenspace, build a light-rail transit system in the longer term, and generate economic development that will revitalize Atlanta's inner-city neighborhoods.

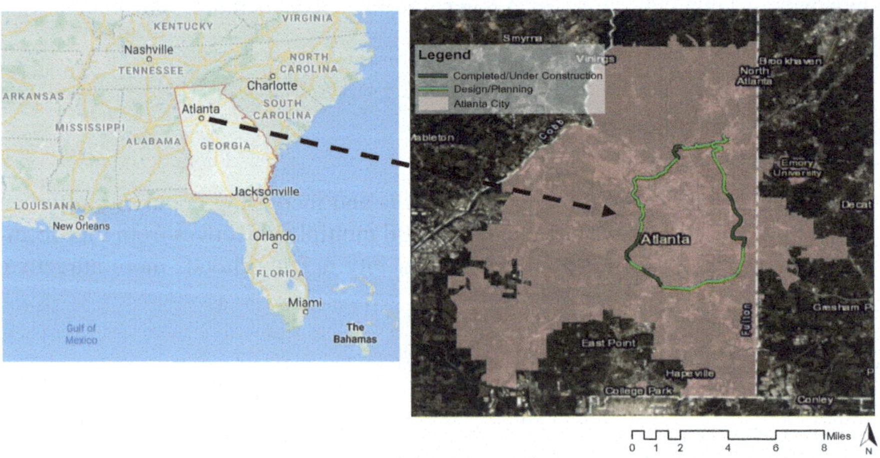


**Fig. 1** Map showing city of Atlanta in Georgia (left), and the Beltline (right). *Source* Google maps 2019, ArcGIS Online 2019 (Esri, Digital globe, GeoEye, Earthstar Geographics, USDA, USGS, AeroGRID, IGN, and the GIS User community)

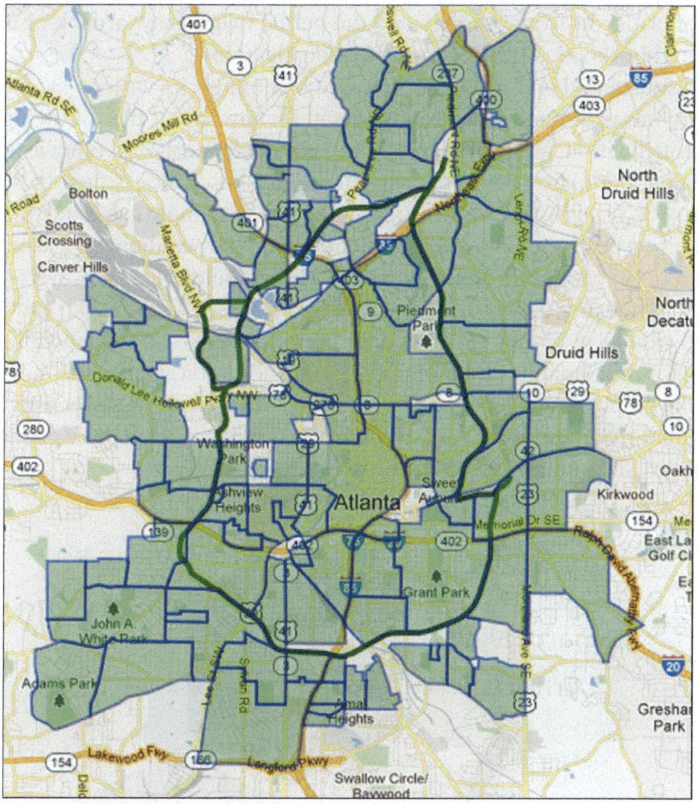


**Fig. 2** Map of neighborhoods impacted by the Atlanta Beltline. *Source* Online GIS Map, Atlanta Beltline (accessed from https://beltline.org/map/?highlight=map), Google Maps 2019

The Beltline project was first envisioned in a Georgia Institute of Technology masters' thesis by Mr. Ryan Gravel, which proposed reusing the railroads to promote private market development of industrial areas, revitalize old streetcar neighborhoods surrounding the downtown, increase tax base by economic development on underutilized lands, create public amenities such as parks and trails, and create a diverse environment for communities to live together [1, 2]. In the two decades following the thesis publication and more than a decade since its construction started in 2008, several elements of the Beltline have already been built, with many of its greenways and parks being actively used by residents—including 8.7 miles of permanent trails, 412 acres of parks and greenspaces, 274 acres of remediated brownfields, and diverse affordable housing options. While the Beltline implementation is still in progress, the city is witnessing redevelopment and economic growth in several formerly depressed neighborhoods where the Beltline has been implemented or is planned. While the Beltline has been hailed for its transformative qualities and quality-of-life improvements, yet as common with many urban redevelopment projects, it has attracted its fair share of criticism from scholars and the public over issues of lack of affordable housing, neighborhood displacement and gentrification.

There is substantial and growing literature examining its various facets. Most of the literature has examined its impact in detail on any one of the following aspects: housing, gentrification, public transportation, urban greenways, or other aspects. This book chapter undertakes a comprehensive literature-based assessment of the Atlanta Beltline project, bringing multiple aspects together in a single case study across four smart living perspectives, namely, (1) urban health and wellbeing, (2) smart public transportation, (3) economic development and brownfields clean-up, and (4) managing affordable housing during neighborhood change (see Fig. 3). Our methodology includes an extensive review of secondary literature from the fields, addressing the project from public health, urban planning, transportation, landscape, housing, and economic development perspectives. In addition, GIS analysis using census and other datasets is undertaken proving a snapshot analysis of socio-economic changes in the neighborhoods surrounding the Beltline over time.

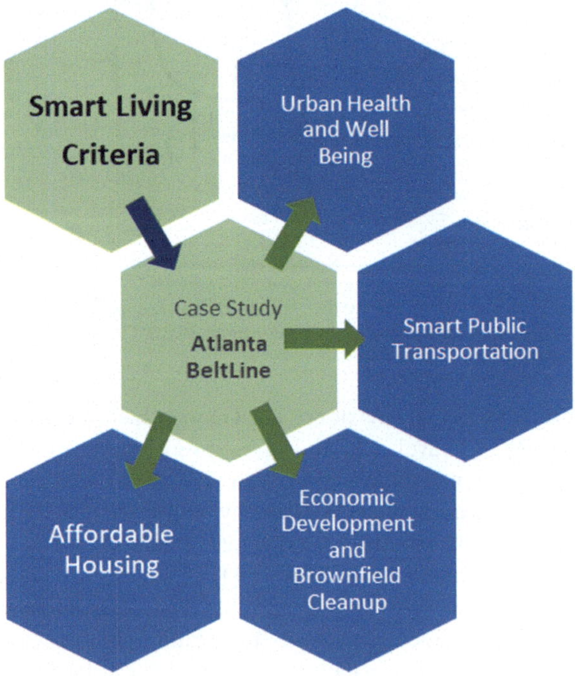


**Fig. 3** Conceptualizing smart living elements for the Atlanta Beltline. *Source* Authors

## 2 Greenspace and Urban Health

The main objective of the Beltline has been to connect the neighborhoods along it through an extensive system of trails and parks. The Beltline is likely to increase Atlanta's greenspace by 40% and add nearly 1300 acres of new parks and gardens and about 700 acres of renovated park space over the next 25 years (www.Beltline.org). An extensive 33 miles of continuous multi-use trail system is being built to connect with over 40 parks (see Fig. 4). Many of the parks are already developed, and over 11 miles of trails have also been completed and opened to public. They not only provide environmental and recreational assets but are also catalytic for new economic development around them. Atlantans who depend on cars waste about 44 h delayed or stuck by traffic every year which can be linked to mental and physical health as well as poor air quality. Further, as per the CDC, 32.5% of adults were obese in the State of Georgia in 2018. The Beltline has the potential to improve Atlanta's health issues as a whole, including obesity. Trails can promote physical activity, improving health, and use of better-connected public transportation in the city can help improve air quality (Fig. 5).

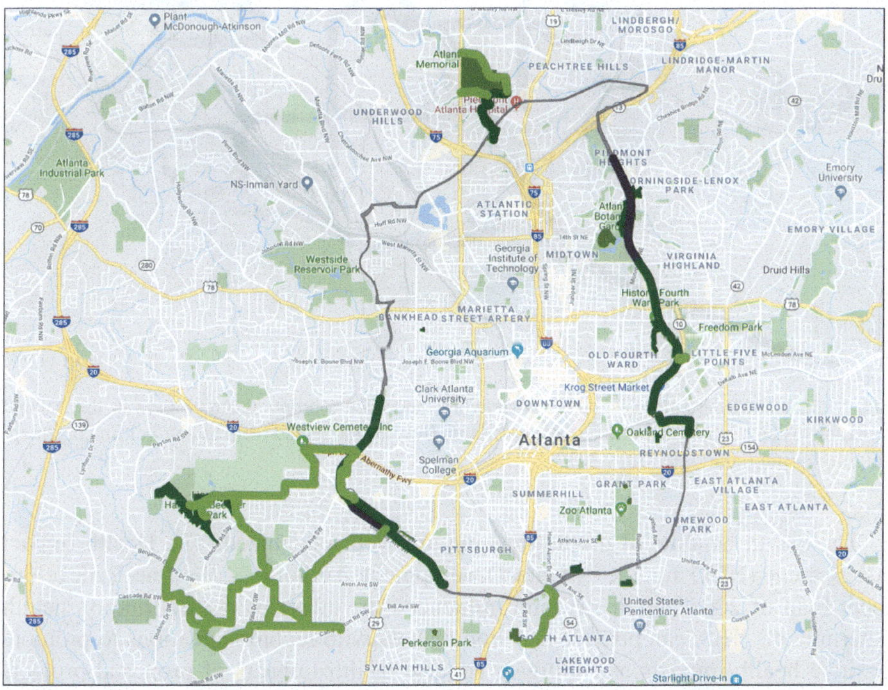


**Fig. 4** Map showing inter-connected system of trails and parks in the Atlanta Beltline project. *Source* Online GIS Map, Atlanta Beltline (accessed from https://beltline.org/map/?highlight=map), Google Maps 2019

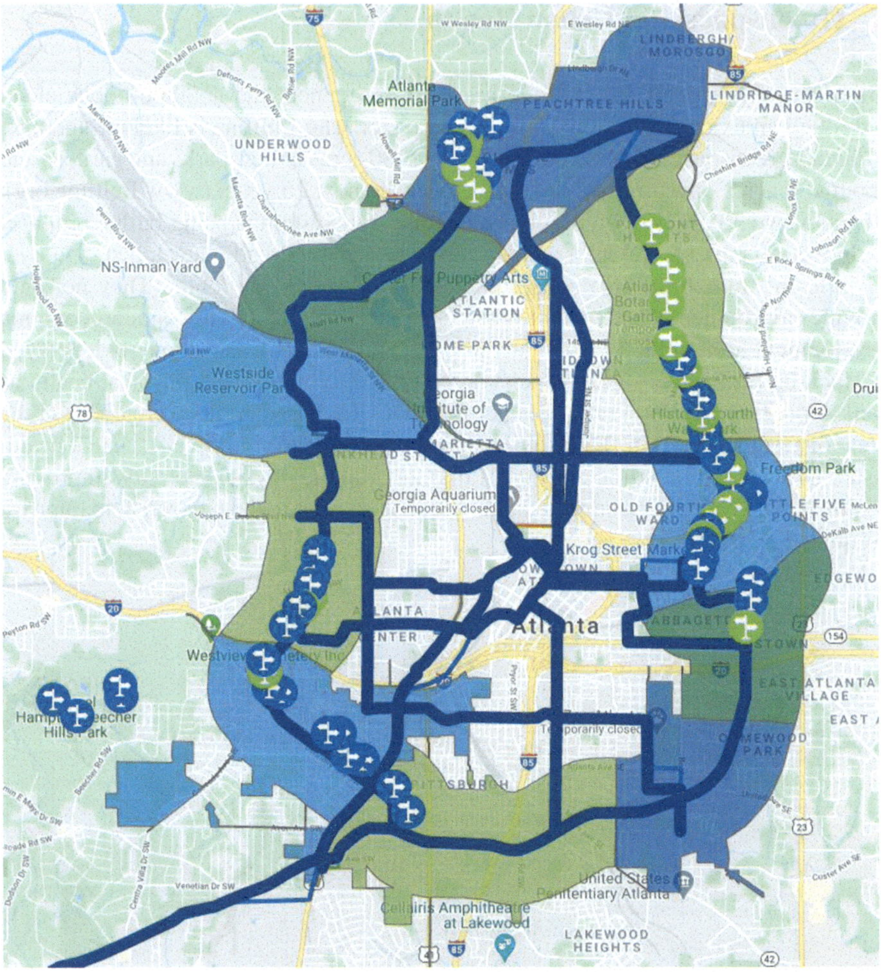


**Fig. 5** Map showing subareas, streetcar network, and access points to greenways in the Atlanta BeltLine Project. *Source* Online GIS Map, Atlanta Beltline (accessed from https://beltline.org/map/?highlight=map), Google Maps 2019

The health benefits of urban greening are well recognized. The Health Impact Assessment (HIA) for the Atlanta Development Authority's Beltline Redevelopment Plan of 2005 anticipated and influenced the effect of the Beltline on health determinants. HIA can provide decision makers with recommendations to promote positive health impacts and mitigate adverse health impacts of proposed projects and policies. Key recommendations included prioritizing construction of trails and greenspace rather than residential and retail, making health an explicit goal in the project, adding public health professionals to decision-boards, increasing connectivity between the Beltline and civic spaces, and ensuring affordable housing. The

HIA was also cited while awarding an additional $7 million for brownfield clean-up and greenspace development [3, 4]. While it is anticipated that the Beltline will increase the number of park acres, it is also expected to increase the population around them. The HIA examined park locations and potential users and accessibility to determine the positive impacts and estimated that an additional 11,000 people will get park access. It suggested to increase access frequency to the trail system, increase accessibility of underserved neighborhoods, create regional creativity, and continuous transition between the transit and trail systems.

Barnes [5] surveyed members of the Beltline Health Proposal Committee (a collaborative effort between the Center for Disease Control and Prevention, Emory University, Georgia Institute of Technology, Georgia State University, the Atlanta Beltline Inc. and the Atlanta Beltline Partnership) and found that members believed that the group was a unique partnership of dedicated professionals sharing a common interest, but faced several challenges such as improving communication, resolving competing interests, and identifying a lead organization. He suggests transitioning to a Health Advisory Group, which can function as a formal organization that will review research proposals, conduct community-based participatory research, leverage funding, and disseminate key health findings.

Attitudes of residents towards Beltline urban greenway support have been studied. Palardy et al. [6] conducted a random survey sampling in the three neighborhoods (Southwest side, Northside, Eastside) and found that residents' support for greenways is not just related to their frequency of its use, but that it is based on more complex perceptions of the economic benefits it provides as well as social and psychological empowerment. They found that psychological empowerment is the strongest predictor for urban greenways. Residents are particularly sensitive to its impact on their pride and self-esteem, and recreation and tourism development have an important impact on the residents' identities. This suggests that variation in greenway design and aesthetics can reflect regional and geographic neighborhood identities and thereby become a tool for social empowerment and neighborhood cohesion. Keith and Boley [7] surveyed the same three neighborhoods and conclude that the residents were overall satisfied with the quality of the trail surface, cleanliness, and natural environment, but were especially dissatisfied with safety (both environmental and physical harm) and suggested adding more emergency call boxes on the trail.

Larson et al. [8] also examined public perceptions to greenway-related benefits in an urban context and found that while environmental benefits (such as cleaner air and water quality) are considered beneficial by people, the cultural benefits (such as economic impacts, social connectivity) or experiential benefits (such as attractive scenery, recreation) were valued and recognized more. The benefit perception also varied across different geographic and socio-demographic contexts. Greenway users were more responsive to experiential benefits regardless of location, because interaction with nature has direct heath impacts on the welfare of users. Keith et al. [9] surveyed users on the Eastside trail and found that exercising and escaping the stress of city life were top motivations, and safety and security were rated as top concerns particularly among women and racial/ethnic minorities. Authors found that many users accessed the trail by foot or bicycle and engaged in a variety of trail-based activities. Cultural benefits linked to social interaction and community connectivity were also widely acknowledged. 32% of visitors to the Eastside Trail were minorities, which also indicated social and nature-based motivations. They suggest that planners can utilize these results to identify strategies for maximizing greenway-related benefits among diverse groups of potential trail users. The location of the greenways and the users' expectations of the benefits needs to be considered by the urban planners and park managers to maximize the utilization of these public amenities targeted for a diverse range of potential users. These greenways offer benefits ranging from stress reduction, socializing to nature-related outdoor experience, along with providing experiential and cultural benefits which as suggested by the research are the most important benefits that the trail visitors associate with. Greenways would be able to improve the urban quality-of-life of a diverse population if the planners can effectively connect the functions and benefits of the trail.

However, evidence of actual health improvements as a consequence of the Beltline seems to be lacking. Dai et al. [10] studied geographic disparities in social determinants of health around the Beltline using an Urban Health Index (UHI) approach for the years 2000 and 2010. They found that there is a gradual improvement in the overall social determinants experienced by Atlanta as shown by a narrow disparity ratio and increase of average UHI. However, they suggest that these improvements may have occurred due to gentrification, leading to displacement of disadvantaged groups, rather than due to actual amelioration of disparities. The Southwest neighborhoods that had low social determinants of health deteriorated more as compared to the Northern neighborhoods of Atlanta. Their research highlights that with revitalization efforts in a city, disparities continue to persist. Tyler [11] examined the birth outcome data and social indicator variables for the pre-Beltline (2002–07) and post-Beltline (2008–12) periods and found that overall there was an improvement trend in birth outcomes, but only a decrease in premature live births in exposed areas and fetal mortality in unexposed areas were statistically significant. Overall, they did not find significant improvements in birth outcome with proximity to the Beltline. Similarly, Fischer [12] found that exposure to traffic-related air pollution was generally the same on and off the Beltline, and there were no noticeable differences. This is contrary to the estimate of the health impact assessment that air pollution in Atlanta will decrease by 2030 due to redevelopment of residential areas and improved transit. However, since the development of transit has not yet started along the Beltline, full impacts in air quality improvements are yet to be seen.

## 3 Urban Transportation

Atlanta's obsession with the traffic perhaps only matches Los Angeles among comparable American Cities. Atlanta has prioritized auto-dependency and sprawl, and has had some of the worst transportation policies as compared to other cities [13]. The Atlanta Beltline will connect 45 Atlanta neighborhoods using a 22-mile loop of trails, parks, and eventually a streetcar system that will be built on existing abandoned railroad tracks. At the core, the Atlanta Beltline is a transit project. A regional light-rail transit system is proposed to be developed in the long term, utilizing the old and defunct railroads to implement a new regional transit that would connect Atlanta's neighborhoods (Fig. 4). Ironically, these railroads of the Atlanta Beltline are symbolic of segregation of the city by race as it once served [13]. The 22-mile Beltline streetcar corridor will be part of a 50-mile integrated streetcar network. This system will also connect to the downtown and the crosstown lines as well as MARTA stations across the city (www.Beltline.org). The Beltline's Health Impact Assessment [4] recommended that transportation improvements should complement alternative travel modes (including MARTA) and pedestrian safety routes improvements in rapidly growing areas, ensure safe routes to school within the Beltline to encourage children to walk and bike to school, implement a coordinated fare and scheduled system for existing and new transit, and people-oriented priorities to address needs of pedestrians, bikers, and transit users in land use regulations. It also recommended access to the Beltline transit system, and new housing and population growth need to be focused near transit stops.

Two major efforts have advanced transit on the Atlanta Beltline: (1) City voters overwhelmingly passed a $0.05 sales tax in 2016, which will fund 16 MARTA projects, including approximately $1.2 billion in funding for Atlanta transit projects. This plan improves connectivity, accessibility, and mobility, and will also bring about significant economic development opportunities across the city–region. (2) Federal NEPA studies also support the Atlanta Streetcar System Plan (SSP) which focuses on connecting neighborhoods to key destinations and activity centers, and establishes criteria for advancing streetcar corridors by evaluating project readiness, equity, value, funding, and ridership. The SSP includes 50 miles of new streetcar routes and 12 miles of connected transit, which includes investment in other forms of public transportation to provide additional connectivity to the streetcar system.

The Atlanta Beltline project has the potential to improve connectivity for pedestrians and cyclists. Cardoni [14] examines the potential of the Beltline to strengthen integrated bike and pedestrian network connectivity using data from a smartphone app, Cycle Atlanta, and surveys. In 2012, the Atlanta bike network was 60 miles long but had little connectivity between them as they were far apart and few. The author identifies several opportunities to improve pedestrian and bike network connectivity which can have a high impact on the overall city transportation, and recommends that Atlanta should explore guidelines for future development on how streets can be redesigned with critical street connections and large bicycle network facilities in order to integrate the Beltline to the city fabric successfully. Kanchik [15] examined select features of the built environment that are related to physical activity using walkability data collected for the Atlanta Beltline. The author carried out a micro-scale analysis of pedestrian walkability features, and found that Beltline segments with a bus stop showed the highest presence of active people (26.3%), followed by segments that had broken/boarded windows/vacant buildings/homes at 21.21%. Surprisingly, streets with shady trees had the lowest presence for active people at 17.99%.

Recent blogposts on *Curbed Atlanta* give some insight into public opinion related to the Beltline. Users are also raising safety concerns over allowing the use of e-scooters. While earlier the Beltline was known for restricting motorized vehicles, now hundreds of e-scooters zoom pass pedestrians and weave in and out of traffic.[1] The city recently passed an ordinance to create a "reduced speed zone" which limits their speed to maximum 8 miles per hour during the busiest hours, weekends and holidays, and implement a "dockless vehicle etiquette" to prevent people from parking them anywhere on the trails.[2]

However, people are much more concerned that transit is taking a backseat in the Beltline development that seems to be more focused on parks and trails. They are frustrated that it might be another 20 years before light-rail tracks are laid, now that the Beltline concept is already 20 years old without any tracks being built. They feel as if 'special interest projects' have been prioritized over transit, even though transit was one of the highlights of the Beltline in the first place when people voted for the sales tax increase.[3] A march was organized in the Historic Fourth Ward, and the activist group Beltline Rail Now! (BRN) believes the rail line will be fully built by 2030.[4] Metro Atlanta's new regional transit authority called The ATL released a list of 76 projects related to expanding the transit system in the city. Building light rail transit for all of the 22-mile Beltline is included in that list, along with MARTA commuter rail lines and bus rapid transit routes in several counties. It is estimated that $15 billion is required to implement every project, but it is debatable if that will actually happen.[5] BRN members have been advocating with the Atlanta Regional Commission to speed up the rail line process, ensure that rail along the entire Beltline remains on the agenda, prioritize transit investment in areas that need connectivity and economic growth rather than affluent areas, and demand better public engagement and transparency by MARTA.[6] An online poll showed 73% support for the transit component for the Beltline, and Ryan Gravel, credited with the original Beltline

---

[1]"Do E-scooters disrupt Atlanta's vision for the Beltline?"' dated 8/23/19. Online resource https://www.ajc.com/news/scooters-disrupt-atlanta-vision-for-the-Beltline/8FCEZsvOUSRHgsGAbfFw6H/#.

[2]"New city program aims to crack down on Atlanta Beltline e-scooter behavior" dated 6/20/19. Online resource https://atlanta.curbed.com/2019/6/20/18692756/slow-e-scooters-8-mph-Beltline-eastside-trail.

[3]"MARTA: Atlantans don't want just any transit on Beltline—they want light rail" dated 4/10/18. Online resource https://atlanta.curbed.com/2018/4/10/17214302/marta-atlanta-Beltline-transit-rail.

[4]"Beltline march planned next month to call for implementation of transit, pronto" dated 9/23/19. Online resource https://atlanta.curbed.com/2019/9/23/20880356/atlanta-Beltline-march-rail-transit.

[5]"Ambitious project list unveiled for metro Atlanta's regional transit plan" dated 9/26/19 online resource https://atlanta.curbed.com/2019/9/26/20885049/project-list-metro-atl-board-rail-marta.

[6]"Beltline rail advocates to MARTA: We want better transparency, better public engagement" dated 7/12/18 Online resource https://atlanta.curbed.com/2018/7/12/17564106/Beltline-rail-marta-tsplost-transit-expansion.

idea in 1999 has also called for transit progress.[7] These efforts have led to over $570 million (of $2.7 billion) from MARTA to be used for 15 miles of light rail tracks,[8] which is an increase from the 7 miles of Beltline transit earlier approved but considered insufficient.[9] In-spite of these efforts, it remains to be seen how long it will take for the Beltline transit to be fully built.

## 4 Brownfields Clean-Up and Economic Development

The clean-up of urban brownfield sites for economic redevelopment has been a popular strategy in old industrial cities of the western world since the 1980s. The impetus behind brownfields clean-up is to create more land in inner-city areas for commercial and residential uses, as opposed to pushing urban development further into greenfield areas outside the city. While brownfield clean-up programs aimed to improve environmental conditions in unused and contaminated industrial sites and create land in inner-cities for revenue-generating urban uses, improvements in quality-of-life of urban residents through urban greening projects are widely practiced in brownfield redevelopment projects. By the 1990s, the majority of cities in North America and Europe focused on comprehensive strategies for brownfield redevelopment with funding from federal or national governments, as well as state and local governments. Examples of these successful brownfield redevelopment projects can be seen in Toronto [16], Pittsburgh [17], New York [18], Chicago [19], Atlanta [20], London [21], Berlin [22] and other major cities. Most of the brownfield redevelopment projects from these cities, on one hand, achieved overall success in transforming underused and derelict land parcels, creating more spaces for employment generation, innovative urban design, and place making ideas to create a new image and sense of place, and increasing up-scale residential and entertainment facilities. On the other hand, they faced criticism for instigating gentrification and the displacement of existing low-income and vulnerable populations. The challenge, thus, remained on how to redevelop brownfield without compromising on social sustainability indicators.

In more recent times, the concept of brownfield redevelopment is applied to create more integrated spaces having a mix of residential, affordable housing, commercial, and office areas as well as green spaces inter-connected with trails and natural areas. Such concepts are aligned with overarching sustainable goals of promoting smart

[7] "Poll: How important is a transit component for the Atlanta Beltline's future?" Dated 3/21/18. Online resource https://atlanta.curbed.com/2018/3/21/17148272/transit-atlanta-Beltlines-future-ryan-gravel.

[8] "Beltline transit advocates lobby ARC for help with trailside rail" dated 4/9/19 and 10/5/18. Online resources https://atlanta.curbed.com/2019/4/9/18301839/Beltline-transit-arc-light-rail-streetcar and https://atlanta.curbed.com/2018/10/5/17940638/Beltline-transit-advocates-more-marta-board-revised-approved.

[9] "MARTA transit plan includes 21 miles of light rail" dated 5/15/18. Online resource https://www.ajc.com/news/local-govt-politics/atlanta-transit-plan-includes-miles-light-rail/i82gtczJk7826JemL087rJ/.

living in the twenty-first century. Our case study of the Atlanta Beltline corridor is no exception to this model of brownfield redevelopment that aims to transform under-utilized land parcels and overall promote smart living at the regional scale.

The Beltline project envisioned an investment of about $20 billion over 25 years of time to bring about significant new businesses, jobs, and other economic development opportunities in the redeveloped areas. In addition to billions of dollars of investment, the Beltline is expected to create 30,000 permanent jobs and 48,000 construction jobs over the years that can further boost the local economy.[10] The purpose of the Beltline project was therefore much more than just providing amenity or a trail system, rather stimulating more holistic economic growth and development opportunities in the inner core of the city and creating more attractive mixed-use communities with better quality-of-life. Public perception studies also supported how economic and cultural benefits of greenways are valued and recognized more by residents than only the health and environmental benefits [8]. To promote economic development opportunities, redevelopment of old industrial and brownfield sites became a prime focus in the Beltline project.

Much of the area along the Atlanta Beltline project intersects with industrial lands and under-utilized brownfield parcels. Since 2005, more than $1 billion has been invested for new private developments that also involved transformation of many brownfield sites (www.Beltline.org). For this transformation, a tax increment finance district was created, implemented in the Atlanta Beltline as Tax Allocation District (TAD, see Fig. 6). TAD was expected to leverage the increase in property taxes paid by the landowners as the consequent of redevelopment and would eventually be used for covering the costs of public amenity. TAD served as an instrument to help fund its own cost by capturing the value that Atlanta Beltline will be generating over time. The investment on the renovations of infrastructures, most visibly being Gargantuan Sears and Roebuck building into Ponce City Market among others, really helped reposition Atlanta's market conditions. Through the investment on these types of infrastructures, the older areas of the city including the obsolete districts were revitalized [2].

This effort of brownfield redevelopment succeeded over time with joint efforts and collaborations between multiple stakeholders: City of Atlanta's Brownfields Program, the Environmental Protection Agency (EPA), City of Atlanta's Office of Planning, Georgia Tech, and Invest Atlanta. An environmental evaluation study of the Beltline corridor in 2005 provided an inventory of 1100 acres of land for brownfield remediation. Further comprehensive analysis of 3200 acres of brownfield sites was initiated in 2010 with brownfield planning grants provided by the EPA. Prime properties, such as the former State Farmers Market and Allene Avenue Urban Farm site, were identified for redevelopment, and land acquisition started shortly after that. In 2011, $850,000 was allocated by the City of Atlanta's Brownfields Revolving Loan Fund (RLF) to facilitate quantifying levels of pollutants and contaminants from past industrial activities along the future Eastside Trail of the Beltline and cleaning them

[10]"Atlanta Beltline Tax Allocation District Feasibility Study." EDAW. 2005.

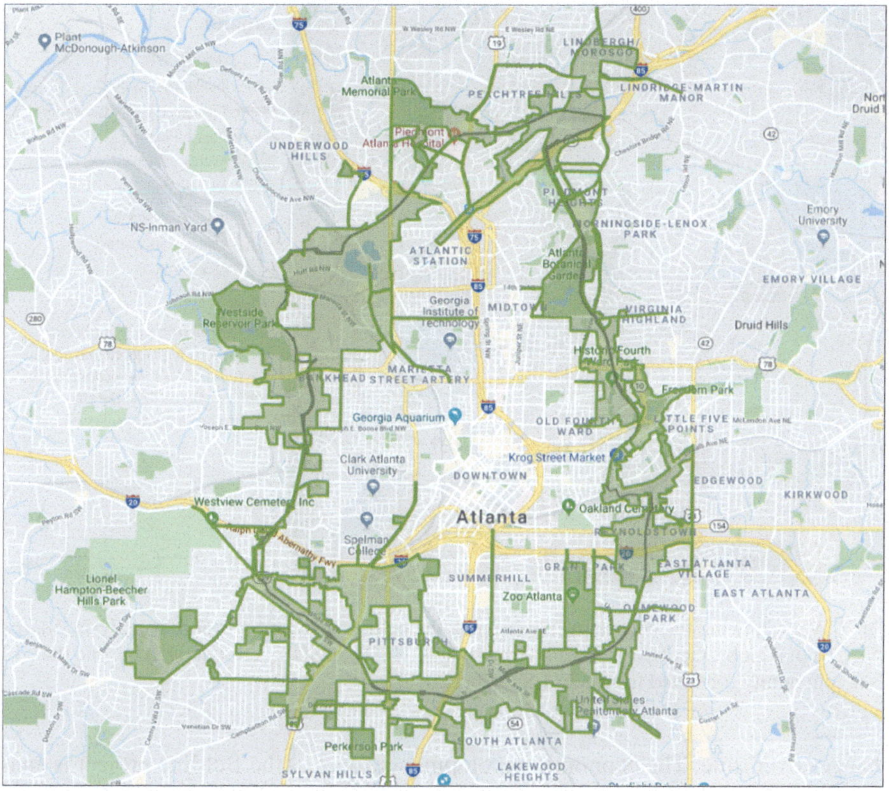


**Fig. 6** Map showing tax allocation districts in the Atlanta Beltline project. *Source* Online GIS Map, Atlanta Beltline (accessed from https://beltline.org/map/?highlight=map), Google Maps 2019

up before trail construction started. Parts of the Westside Trail construction started in 2015 with similar efforts of environmental clean ups (Beltline.org).

Following many successful projects of brownfield clean-ups, trails, parks, greenways, and mixed-use development projects started to spur in the old industrial sites resulting in local economic growth and development. To access the impacts of such brownfield cleanup and economic development projects along the Atlanta Beltline, we selected the census tracts adjoining the Beltline (see Fig. 7). Socio-economic datasets for these census tracts are collected from U.S. Census Bureau for two time-periods: (1) Year 2000—before the implementation of the Beltline project, and (2) Year 2017—the most recent time period for which data is available. This time span captures more than a decade of the Beltline project implementation phase, and as a result generates some key insights on socio-economic changes that have been witnessed in neighborhoods adjoining the Beltline. We extensively mapped these datasets using ArcMap 10.6 and socialexplorer.com to assess the socio-economic

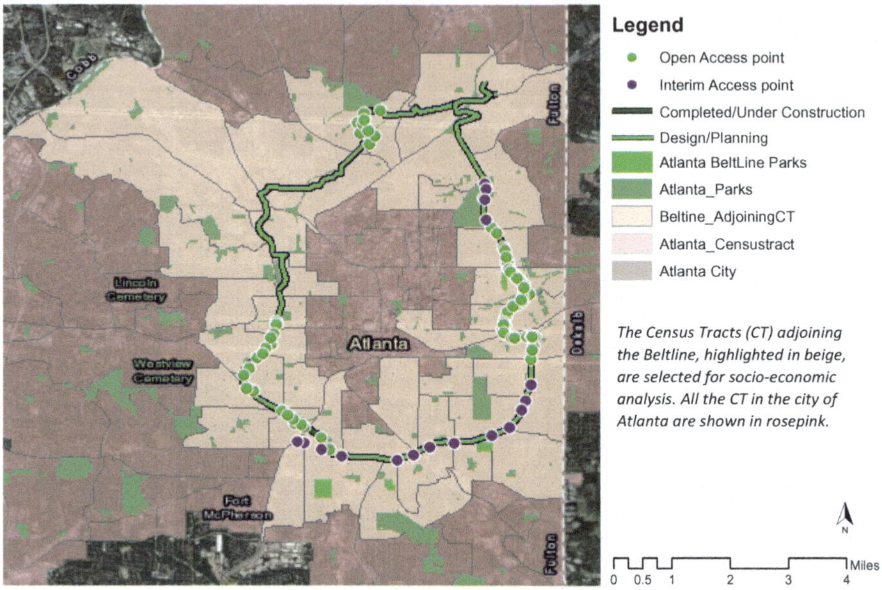


**Fig. 7** Map showing the census tracts adjoining the beltline for socio-economic analysis. *Source* ESRI, Digital globe, GeoEye, Earthstar Geographics, USDA, USGS, AeroGRID, IGN, and the GIS User community (obtained from ArcGIS Online 2019)

changes over time. The economic development aspect of the Beltline project is analyzed based on changes in four indicators:(i) Income and Educational Level, (ii) Housing Market, (iii) Racial distribution, and (iv) Poverty level.

## *4.1 Changes in Income and Educational Level*

Income and Educational level across the census tracts (CT) adjoining the Beltline witnessed different levels of variations in between 2000 and 2017. We collected data on Median Household Income and % of population with a bachelor's degree or more to analyze this. Some CT did not witness much change in these levels while some CT witnessed significant changes. We identified 10 census tracts, where there has been some significant increase in income and educational levels (see Figs. 7 and 8 and Table 1). It is also observed that these are mostly the CT where the Beltline is either completed or under construction, as well as the majority of access points to the Beltine are already open to the public in these areas. The CT where the Beltline is in the planning stage did not witness such high levels of income and educational attainment changes.

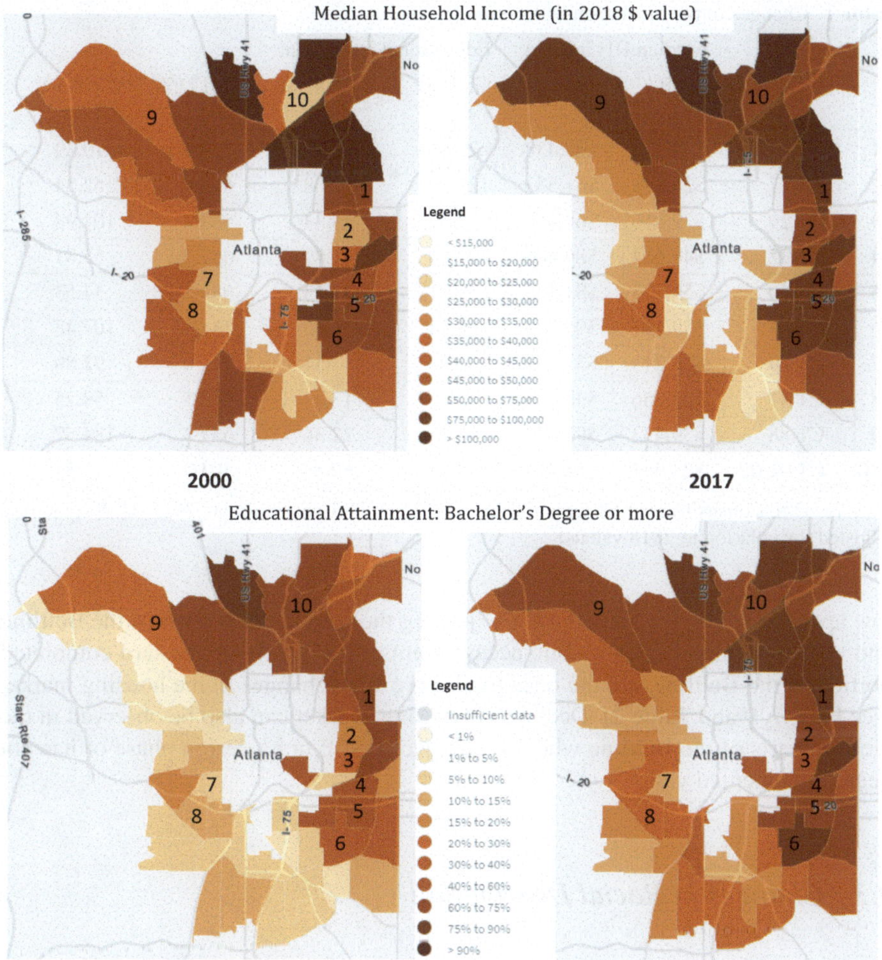


**Fig. 8** Changes in income and educational levels along the Beltline. *Source* Authors, data obtained from Socialexplorer.com

## 4.2 Changes in Housing Market

Housing Market has drastically changed between 2000 and 2017 in those 10 CT as well, that observed a significant increase in income and educational levels. The 10 census tracts we identified mostly witnessed a drastic increase in two indicators—median housing value as well as median gross rent. Census Tract 17 specifically witnessed a 193% increase in housing values from $129,701 in 2000 to $381,068 in 2017, after all values were adjusted for inflation to 2018 $ values. Gross rent in this CT also shot up by 126% from $623 to $1411. A few of the CT that observed decline in the housing market, e.g. Census Tract 14, only witnessed moderate or

**Table 1** Changes in the income and education levels

| | | Median HH income | | | Educational attainment | | |
|---|---|---|---|---|---|---|---|
| | | 2000 | 2017 | Gr. Rate (%) | 2000 (%) | 2017 (%) | Gr. Rate (%) |
| 1 | CT 14 | $69,714 | $88,438 | 26.86 | 69.88 | 77.14 | 10.39 |
| 2 | CT 17 | $27,071 | $61,533 | 127.30 | 17.08 | 61.87 | 262.24 |
| 3 | CT 29 | $39,816 | $64,650 | 62.37 | 28.34 | 58.09 | 104.98 |
| 4 | CT 32 | $48,684 | $86,406 | 77.48 | 41.89 | 65.27 | 55.81 |
| 5 | CT 50 | $59,733 | $81,182 | 35.91 | 44.53 | 68.81 | 54.53 |
| 6 | CT 53 | $59,191 | $95,734 | 61.74 | 36.65 | 76.04 | 107.48 |
| 7 | CT 39 | $28,213 | $37,056 | 31.34 | 7.31 | 14.10 | 92.89 |
| 8 | CT 41 | $32,856 | $39,458 | 20.09 | 13.49 | 21.55 | 59.75 |
| 9 | CT 88 | $39,433 | $97,853 | 148.15 | 22.36 | 63.63 | 184.57 |
| 10 | CT 91.02 | $24,963 | $55,495 | 122.31 | 62.54 | 70.99 | 13.51 |

*Sources* U.S. Census Bureau Decennial Census 2000, ACS 5-year estimate 2017 (all $ values are adjusted for inflation to 2018 values)

low levels of decline. Again, the CT adjoining the completed portion of the Beltline and with more numbers of public access points, specifically the eastern completed portion of the Beltline, are the ones to observe such changes in the housing market (see Figs. 7, 9 and Table 2). Decline in Housing Market can also be observed in the western parts of the Beltline where the project is still in its design phase or has not been implemented yet.

## 4.3 Changes in Racial Distribution

The Beltline project runs through many of the African American majority neighborhoods that traditionally were underinvested and in poor conditions. Hence, many of these neighborhoods had a high concentration of African American population, and lower concentrations of White or other racial groups. We collected data on % of African Americans and % of White population to analyze the changes in racial makeup in these census tracts (CT). The majority of the predominantly African American neighborhoods were in the western portion of the Beltline in 2000, while the CT along the eastern portion of the Beltline also had a significant amount of African American population. Many of these started witnessing a declining share of the African American population after the creation and implementation of the Beltline project (see Fig. 10 and Table 3). This trend is also correlated with an increasing share of the White population in the same areas. Not only the share but also the absolute number of African American population has declined, and White population has increased respectively. Census Tract 17, for example, again witnessed a decline of

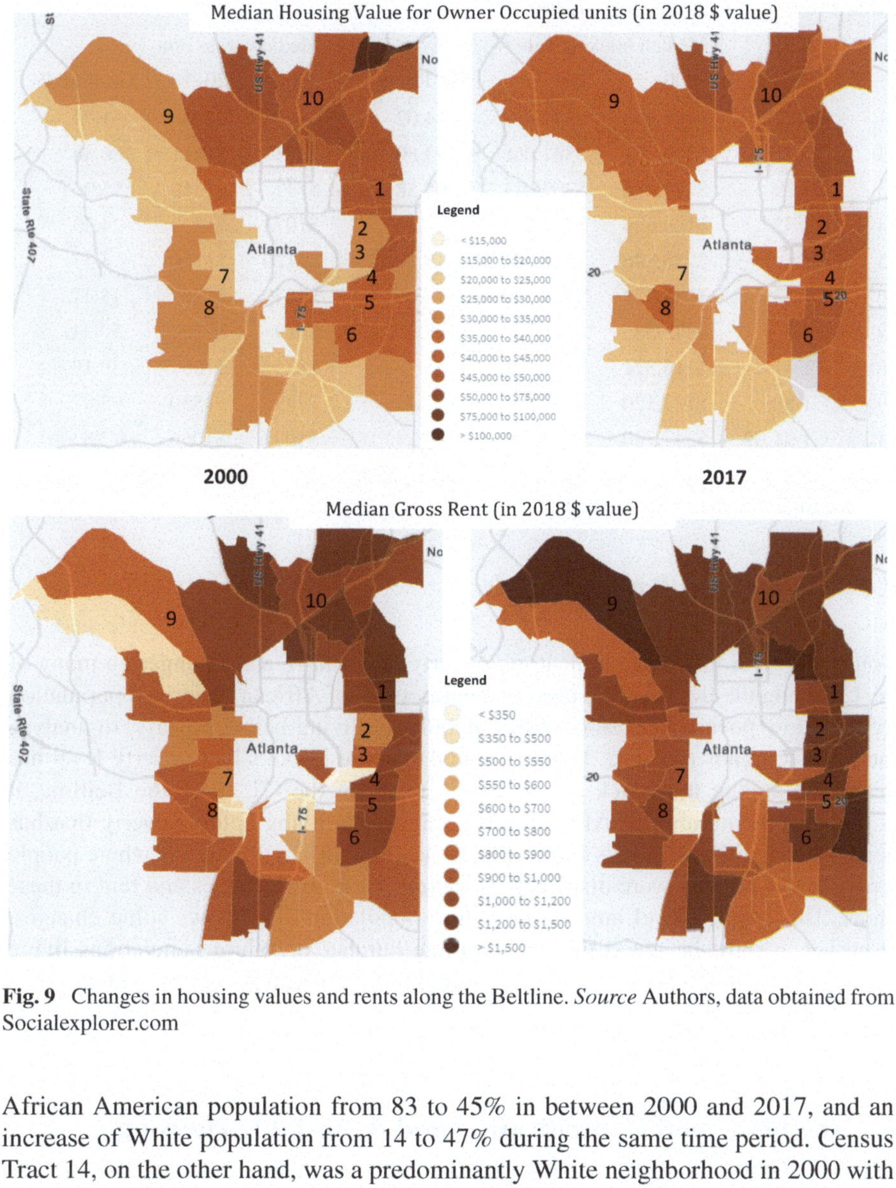


**Fig. 9** Changes in housing values and rents along the Beltline. *Source* Authors, data obtained from Socialexplorer.com

African American population from 83 to 45% in between 2000 and 2017, and an increase of White population from 14 to 47% during the same time period. Census Tract 14, on the other hand, was a predominantly White neighborhood in 2000 with only 12% African American population and 84% White population. Neighborhoods such as this witnessed even higher decline of African American population in absolute numbers. This indicates a strong trend of ongoing gentrification and changing racial distribution of these areas close to the Beltline project. The western portion of the Beltline, still in its planning phase, has not yet witnessed such a dramatic change of its racial makeup.

**Table 2** Changes in the housing market

| | | Median housing value | | | Median gross rent | | |
|---|---|---|---|---|---|---|---|
| | | 2000 | 2017 | Gr. Rate (%) | 2000 | 2017 | Gr. Rate (%) |
| 1 | CT 14 | $462,506 | $397,663 | −14.02 | $1268 | $1246 | −1.74 |
| 2 | CT 17 | $129,701 | $381,068 | 193.80 | $623 | $1411 | 126.48 |
| 3 | CT 29 | $147,394 | $303,011 | 105.58 | $747 | $1310 | 75.37 |
| 4 | CT 32 | $188,629 | $234,275 | 24.20 | $1073 | $1333 | 24.23 |
| 5 | CT 50 | $280,896 | $336,508 | 19.80 | $1120 | $1160 | 3.57 |
| 6 | CT 53 | $281,773 | $329,849 | 17.06 | $1070 | $1267 | 18.41 |
| 7 | CT 39 | $87,588 | $62,794 | −28.31 | $619 | $930 | 50.24 |
| 8 | CT 41 | $115,371 | $179,676 | 55.74 | $826 | $1067 | 29.18 |
| 9 | CT 88 | $123,120 | $292,050 | 137.21 | $779 | $1550 | 98.97 |
| 10 | CT 91.02 | $347,720 | $571,090 | 64.24 | $1097 | $1066 | −2.83 |

*Sources* U.S. Census Bureau Decennial Census 2000, ACS 5-year estimate 2017 (all $ values are adjusted for inflation to 2018 values)

### *4.4 Changes in Poverty Levels*

Consistent with the above indicators, poverty levels have also changed in many of the CT along the Beltline. We used two datasets, % of African American population living below poverty and % of White population living below poverty, to analyze the poverty trends (see Fig. 11 and Table 4). While it looks like poverty level has decreased among the Black population in many of the CT along the Beltline, it is the absolute number of African American people living below poverty that has decreased significantly. This can be indicative of gentrification trends where people living below poverty were displaced by increasing housing values and rent in these areas. The poverty level among the White population also shows some changes. However, it must be noted that the absolute number of White populations living below poverty levels are negligible compared to the African American population in these CT.

## 5 Neighborhood Gentrification and Affordable Housing

Strategic infrastructure investment has emerged in recent years as one of the key catalysts for urban revitalization among global cities in developed countries [23–26]. A consequence of this vitality is gentrification: the inevitable loss of affordable housing due to rising property values and rent, displacement of low income and minority resident neighborhood populations, who are supplanted by affluent, young professional newcomers who are usually white [27, 28]. Associated with gentrification-induced change is place-making that includes desirable facilities such

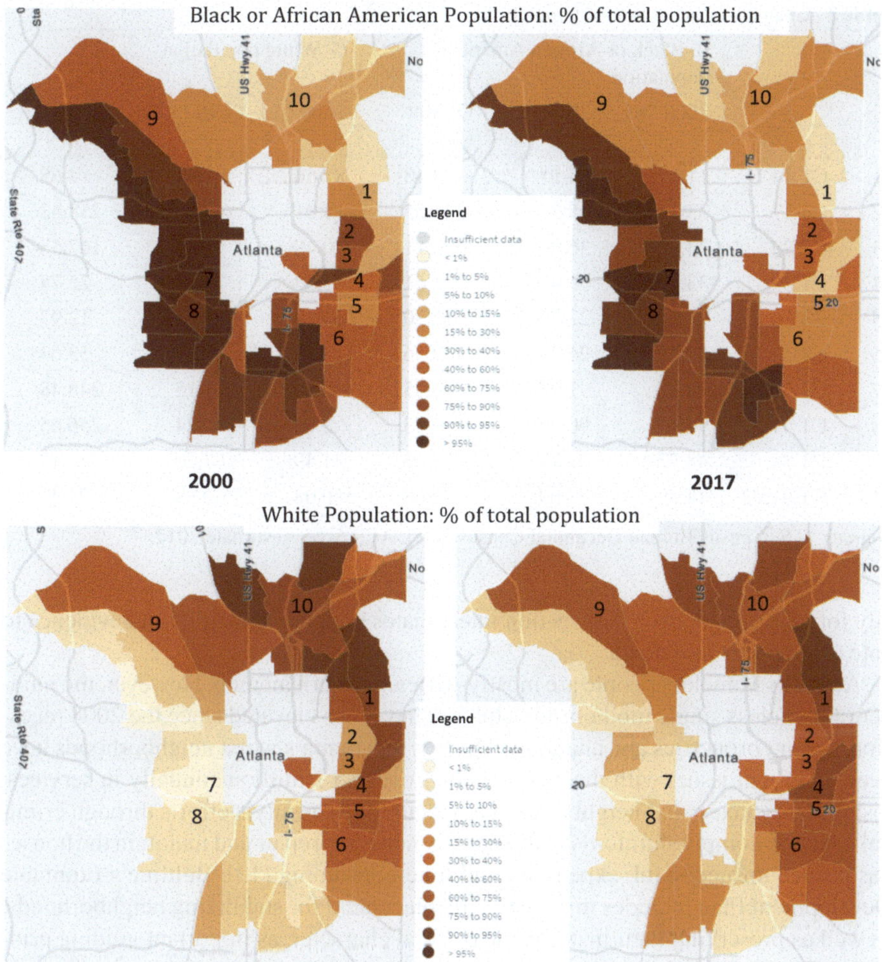


**Fig. 10** Changes in racial mix of population along the Beltline. *Source* Authors, data obtained from Socialexplorer.com

as parks, urban trails, improved public transit and bike lanes—all positive attributes that enhance livability in urban areas. What confounds planners and decision makers is the outcry by longtime residents captured by Saunders (2018) in the caption "*Now that the neighborhood is nice, why do I have to move*?" [28].

This question is at the vortex of urban revitalization intensity along the Atlanta Beltline corridor. As anecdotal evidence began to emerge in the formative years of the Beltline project exhibiting some elements of traditional gentrification characteristic effects—displacement and loss of affordable housing—the city sought to develop programs and strategies to strike a balance for positive transformation [2, 4, 13]. How has this interventionist/proactive approach worked to make Atlanta a livable

**Table 3** Changes in racial mix of populations

| | | % Black or African American population | | | % White population | | |
|---|---|---|---|---|---|---|---|
| | | 2000 (%) | 2017 (%) | Gr. Rate (%) | 2000 (%) | 2017 (%) | Gr. Rate (%) |
| 1 | CT 14 | 11.69 | 3.11 | −73.40 | 83.66 | 87.55 | 4.65 |
| 2 | CT 17 | 83.16 | 45.28 | −45.55 | 14.01 | 47.44 | 238.62 |
| 3 | CT 29 | 73.97 | 36.90 | −50.11 | 20.48 | 54.80 | 167.58 |
| 4 | CT 32 | 33.18 | 9.42 | −71.61 | 62.02 | 78.01 | 25.78 |
| 5 | CT 50 | 27.12 | 22.70 | −16.30 | 64.29 | 72.63 | 12.97 |
| 6 | CT 53 | 52.25 | 24.79 | −52.56 | 43.40 | 66.68 | 53.64 |
| 7 | CT 39 | 98.23 | 97.98 | −0.25 | 0.33 | 1.15 | 248.48 |
| 8 | CT 41 | 94.11 | 90.79 | −3.53 | 2.03 | 7.24 | 256.65 |
| 9 | CT 88 | 48.72 | 21.99 | −54.86 | 41.49 | 67.70 | 63.17 |
| 10 | CT 91.02 | 14.39 | 22.69 | 57.68 | 80.08 | 62.88 | −21.48 |

*Sources* U.S. Census Bureau Decennial Census 2000, ACS 5-year estimate 2017

city for all its residents? This section interrogates both the strategies and evidence to date as it relates to Atlanta.

Atlanta's homeless people are mostly African-American men. However, the number of homeless single mothers and children have also elevated since the 2008 recession. Vacant properties and abandoned homes in Atlanta's worst neighborhoods have been an urgent issue, with the city spending over $5.7 million annually in services, lost property taxes, and neighborhood depreciation. The city's efforts through criminal courts to compel slumlords and absentee owners to repair and maintain the houses has proven unsuccessful, expensive, and time-consuming [13]. Beltline's Equitable Development Plan includes minimizing displacement for stabilizing neighborhoods, as well as preserving their historic and cultural character, as important guiding principles. The Beltline has been critiqued for giving too much emphasis to economic development and not having enough focus on affordable housing. Approximately 560 units of affordable housing have been supported by the Beltline. About $15–20 million of Tax Allocation District revenue funding commitment has been made for additional affordable housing construction, and 265 additional units are planned for 2017 (www.Beltline.org).

Low income communities had concerns over economic displacement but nevertheless supported the Atlanta Beltline from the beginning. The project brought in financial tools that could help counterbalance these financial pressures. The Tax Allocation District (TAD), defined with the inclusion of housing subsidies, was one of the key attempts to tackle affordability, and the legislation allocated 15% of TAD revenues towards housing affordability. This helped in funding grants for preserving affordable housing and to assist in down payments [2]. Neighborhood conservation was one of the Atlanta Beltline's core concerns as the neighborhoods were wary for

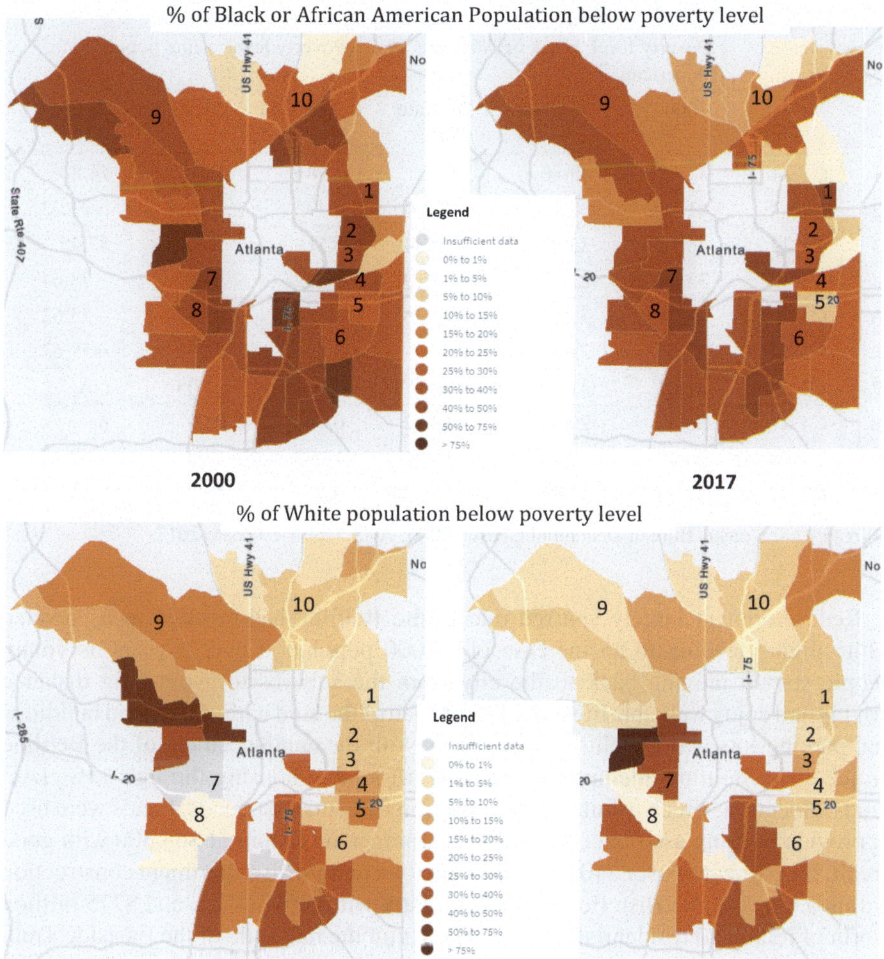


**Fig. 11** Changes in poverty level along the Beltline. *Source* Authors, data obtained from Socialexplorer.com

unfavorable transformation and were highly aware about protecting themselves, not from changes themselves, but to "protect their opportunity to leverage that change in their advantage," [2]. The project's original proposal promised them that single family housing areas would be protected from high-density developments through zoning, and that neighborhood revitalization would happen through infill housing on vacant lands and commercial and cultural districts in appropriate areas [2]. The Health Impact Assessment (HIA) recommended that housing should be built to suit the needs of diverse groups of the population and encourage diversity, and to reduce the impact of increasing property tax assessment on lower income families to prevent displacement issues by establishing policies and programs [4].

**Table 4** Changes in poverty levels

| | | Poverty level: black or African American | | | Poverty level: white population | | |
|---|---|---|---|---|---|---|---|
| | | 2000 (%) | 2017 (%) | Gr. Rate (%) | 2000 (%) | 2017 (%) | Gr. Rate (%) |
| 1 | CT 14 | 32.15 | 49.30 | 53.34 | 2.98 | 5.21 | 74.83 |
| 2 | CT 17 | 41.34 | 28.11 | −32.00 | 9.49 | 2.56 | −73.02 |
| 3 | CT 29 | 36.16 | 25.37 | −29.84 | 6.51 | 10.25 | 57.45 |
| 4 | CT 32 | 33.69 | 22.17 | −34.19 | 14.42 | 3.08 | −78.64 |
| 5 | CT 50 | 24.21 | 5.48 | −77.36 | 17.55 | 8.14 | −53.62 |
| 6 | CT 53 | 27.61 | 29.69 | 7.53 | 8.94 | 3.61 | −59.62 |
| 7 | CT 39 | 41.42 | 41.23 | −0.46 | NA | 31.03 | NA |
| 8 | CT 41 | 36.57 | 29.98 | −18.02 | 0.00 | 0.00 | na |
| 9 | CT 88 | 36.25 | 29.63 | −18.26 | 18.83 | 1.99 | −89.43 |
| 10 | CT 91.02 | 21.33 | 15.94 | −25.27 | 9.57 | 8.07 | −15.67 |

*Sources* U.S. Census Bureau Decennial Census 2000, ACS 5-year estimate 2017

Redevelopment largely spurred due to the Beltline has resulted in a "reverse white flight" leading to an increase of 35,000 population over 5 years as young people started moving back to the city from the suburbs to avoid long distance driving and enjoy walkable lifestyles [13]. Much of the land with abandoned buildings and parking lots was bought by speculators with the announcement of the Beltline project. The Beltline attempted to provide affordable housing through 15% TAD funds allocation, which resulted in $8.8 million from the first bonds. Funds were used to provide housing assistance to families earning a maximum of $68,300 with good credit. Investment of over $400 million went toward high-rise apartment construction within a block of the Historic Fourth Ward Park within two years, and $775 million worth of retail and residential investments within the half mile of the Eastside Trail. An affordable housing ordinance was passed in the spring of 2016 which required for developers receiving public funds to provide 15% of new units to people making 80% of the Area Median Income (AMI) or 10% of the new units to people making 60% of the AMI. A similar inclusionary zoning mandate was introduced for all new multifamily housing in the TAD despite no public money involvement. However, Ryan Gravel and Nathaniel Smith, founders of the Partnership for Southern Equity, resigned from the Atlanta Beltline Partnership Board in September 2016 to protest lack of focus on affordable housing and equity issues, and too much focus on fund-raising [13].

In the early stage of the project, Immergluck [29] examined home sale data between 2003 and 2005 to assess the impact of the Beltline announcement on low-income neighborhoods and found substantial speculation and gentrification as the outcome of redevelopment that resulted in the displacement of low-income households due to TIF. The increase in premiums were higher for homes near lower-income,

Southside parts of the Beltline TIF district, which corresponds to the initial media coverage of the plan. The Anti-Displacement Tax Fund that was created to offset the increase in taxes for homeowners greatly underestimated the number of eligible residents and costs, as it only estimated 400 residents to be eligible for the program [30]. The authors estimated that the program cost over 20 years will increase drastically as compared to initial projection with the increase in the number of eligible homeowners and larger property value appreciation rates.

In their recent research Immergluck and Balan [31] conclude that while the Beltline will most likely be considered a success in increasing both recreation and Atlanta's property tax base, those lower-income residents who live near current Beltline projects are negatively affected economically. The Beltline is bringing in development, craft breweries, restaurants, and luxury apartments, but displacing original communities by triggering sharp increases in home values in low-income and largely African-American communities in the southwest segment. Their study found that median sale prices in some neighborhoods increased 68% from 2011 to 2015. They suggest that in the rest of the project's development, there should be more effort to provide affordable housing options accessible to lower income households and help existing residents remain in these neighborhoods. Our analyses of the changing housing market along the Beltline also support similar trends of gentrification and decrease in availability of affordable housing along the completed portions of the Eastside as well as some parts of the Westside trails (see Fig. 9). The African American communities are also adversely affected by the success of the Beltline projects, whose population has significantly decreased in many neighborhoods (see Figs. 10 and 11 and Table 4).

Understandably, public opinion on the Beltline has been split between both positive and negative. Weber et al. [32] surveyed neighborhoods along the Beltline and found that it was seen to have positive development connotations such as park development, places for outdoor recreation, property value appreciation and neighborhood business development, but it also worsened the negative connotations such as crime, traffic, property taxes, litter and vandalism. Responses varied in different neighborhoods—Northside neighborhoods responded positively to the impacts (recreational benefits) as compared to Southwest neighborhoods (economic development opportunities). Their results also indicate residents' concern over gentrification and crime associated with greenways. Hsieh et al. [33] finds that the prospect of increased property values generally resulted in residents' support for the Beltline, which was seen as a sustainable smart city initiative bringing positive impacts. However, according to the "homevoter hypothesis", public support may be adversely impacted as some people (such as parents of school-going children) might not support the tax incremental financing (TIF) mechanism. TIF would allow Atlanta to divert future increase in tax revenues towards the development of the Beltline from other important sectors such as public education system.

## 6 Housing Affordability

Housing affordability remains a daunting challenge for the Beltline development corridor a decade later, following implementation in 2006. Only 785 affordable units had been provided, a far cry of the 5600 affordable housing targeted[11], prompting Brian McGowan, Beltline CEO to acknowledge a "C grade performance" for the period. Significantly, Beltline brain trust, Gravel, and Paul Morris, original Beltline CEO have since resigned over the housing affordability issue [13]. A key factor for poor performance has been a rise in real estate values. According to the 2017 Georgia Tech Study, the Beltline corridor and its environs experienced a rise in property values between 17.9 and 26.6%. This finding is consistent with our own analysis noted previously (Fig. 9). The undeniable fact is that gentrification is inducing displacement, exclusion and unaffordability, particularly in traditional neighborhoods which have been overwhelmingly African American. Consequently, lack of housing options remains a hotly-contested issue along the Beltline corridor.

To counteract the challenge of housing affordability, the following measures have been either recently implemented or proposed. First, an inclusionary zoning ordinance was promulgated citywide in 2018 requiring developers of all residential rental developments consisting of ten (10) or more new dwelling units to set aside at least the following:

- 10% of their units for incomes at or below 60% of Area Median Income (AMI); or
- 15% of their units for incomes at or below 80% of AMI; or
- Pay a one-time in-lieu fee, to be paid at 15% of AMI, per unit in the sub-area that the developer has chosen to opt-out of, in-lieu of setting aside affordable units (www.Beltline.org).

Secondly, Atlanta's mayor, Keisha Bottoms, recently initiated a multi-pronged city-wide Housing Affordability Action Plan that aims, among other things, to "create/preserve 20,000 affordable homes by 2020, through minimized displacement, and leveraging $1 billion from public, private and non-profit sector" (www.Beltline.org). In addition, the non-profit, Westside Future Fund, initiated an Anti-Displacement Tax Fund, targeting predominantly African American neighborhoods on the Westside (CT 39 and 41) that are yet to experience Beltline development. This is a preemptive strategy designed to protect against speculation while seeking to preserve affordability for homeowners and renters. With a homeownership rate of only 8%, the Anti-Displacement Tax Fund seeks to freeze property taxes of qualifying homes owned by residents making the area median income or below, which may become unaffordable due to rising property values induced by speculative development along the Beltline. For renters, the Fund will purchase properties to boost availability and affordability. A recent study by researchers at Georgia Tech illustrates (see Table 5)

[11] "Atlanta's Beltline, a transformative urban redevelopment, struggles with affordability" dated 10/03/2017. Online resource https://www.curbed.com/2017/10/3/16411354/Beltline-atlanta-affordable-housing-development-high-line.

**Table 5** Potential impact of westside future fund's anti-displacement tax

| Neighborhood | Total residential properties | Homes eligible for fund | Percent of residential HH eligible (%) | WFF payments to eligible participants (2018–2014) | Total change in residential property taxes (2018–2024) |
|---|---|---|---|---|---|
| Ashview heights | 524 | 140 | 27 | $1,116,081 | $3,986,969 |
| AUC | 148 | 20 | 14 | $145,807 | $1,390,661 |
| English avenue | 695 | 194 | 28 | $876,141 | $3,286,260 |
| Vine city | 682 | 206 | 30 | $1,058,258 | $3,291,830 |
| Washington park | 545 | 142 | 26 | $1,016,126 | $4,137,558 |
| Total | 2594 | 702 | 27 | $4,212,414 | $16,093,277 |

*Source* Adapted from Curbed[a], Property Lines
[a]"Atlanta's Beltline, a transformative urban redevelopment, struggles with affordability" dated 10/03/2017. Online resource https://www.curbed.com/2017/10/3/16411354/Beltline-atlanta-affordable-housing-development-high-line

the potential positive impact of the Anti-Displacement Tax Fund strategy.

While there is considerable development in the neighborhoods along the corridor, gentrification is leading to reduction in the affordable housing stock. It remains to be seen whether these renewed efforts will be sufficient enough to counteract development forces that have been experienced to date while delivering the promise of housing affordability. Evidence from other cities that have sponsored similar projects, for example the High Line in New York or Chicago's 606, suggests that the battle may have been lost already.

## 7 Conclusion

Based on a comprehensive literature-based assessment of the Atlanta Beltline project, under implemented since 2008, our study brings multiple aspects and impacts of the project together in one case study. Our assessment across four smart living perspectives, namely, urban health and wellbeing, smart public transportation, economic development and brownfields clean-up, and managing affordable housing during neighborhood change, shows that often the positive values of such large-scale and transformative urban regeneration and greenway programs are not much acknowledged. With various greenway, transportation, housing, and economic development projects already implemented, and an anticipated investment of $20 billion over 25 years of time, the Beltline has significantly spurred transformative urban regeneration in previously neglected areas and brownfields of Atlanta, GA. While considerable success of the Beltline can be asserted in form of financial investment, economic growth, and addition of green spaces and trails; the process has nevertheless been accompanied by gentrification and displacement that disproportionately affected the racial minorities and poor neighborhoods in the recent times. It is also identified that economic, cultural, and experiential outcomes are just as important as health and transportation outcomes.

In case of the Atlanta Beltline, we observed changing priorities from planning implementation perspective, over time the planning projects focused more and more in creating recreational trails and developing urban parks while neglecting other important areas, such as the crucial light rail transit system and affordable housing projects. The inevitable consequences of such shifting priorities have been gentrification and displacement of the urban poor, specifically the minority African-American populations. Our socio-economic analyses of the selected census tracts surrounding the Beltline highlight these trends—increasing housing value and rent, decreasing African-American and increasing White populations, increasing income and educational levels, and changing poverty levels as a result of displacement of the urban poor in between 2000 and 2017. The current policies and programs within the city of Atlanta are not able to address these issues adequately, and further leading to complex social justice issues.

The story of the Atlanta Beltline, thus, resonates quintessential planning challenges witnessed in many contemporary successful projects across the United States—how to promote sustainable urban development goals that can benefit all? The economic cost-benefit analyses of the Beltline unquestionably portrays how a few specific groups of people, in this case the educated population with higher incomes who are also White in majority, benefit more from smart living ideas, such as having access to trails, greenways, parks, jobs and other economic and recreational opportunities that increasingly define the twenty-first century smart living paradigms. The quest for equitable distribution, social justice and the rights to the city, therefore, remains on. The question for future planning endeavors in smart living implementation that needs to be prioritized by policy makers is—how can the different urban sustainable development goals be balanced in these transformative urban projects?

## References

1. Gravel R (1999) Belt line—Atlanta: design of infrastructure as a reflection of public policy. Master's thesis, College of Architecture, Georgia Institute of Technology
2. Gravel R (2016) Where we want to live: reclaiming infrastructure for a new generation of cities. St. Martin's Press, New York
3. Ross CL, de Nie KL, Dannenberg AL, Beck LF, Marcus MJ, Barringer J (2012) Health impact assessment of the Atlanta Beltline. Am J Prev Med 42(3):203–213
4. Ross CL, West H (2007) Atlanta Beltline health impact assessment. Georgia Institute of Technology
5. Barnes B (2011) A multi-vitamin for the built environment: exploring how a multi-sectoral and multi-institutional partnership links health and the Beltline. Unpublished thesis, School of Public Health, Georgia State University

6. Palardy NP, Boley BB, Gaither CJ (2018) Residents and urban greenways: modeling support for the Atlanta Beltline. Landscape Urban Plann 169:250–259
7. Keith SJ, Boley BB (2019) Importance-performance analysis of local resident greenway users: findings from Three Atlanta Beltline Neighborhoods. Urban Forestry Urban Greening 44:1–13
8. Larson LR, Keith SJ, Fernandez M, Hallo JC, Shafer CS, Jennings V (2016) Ecosystem services and urban greenways: what's the public's perspective? Ecosyst Serv 22:111–116
9. Keith SJ, Larson LR, Shafer CS, Hallo JC, Fernandez M (2018) Greenway use and preferences in diverse urban communities: implications for trail design and management. LandscapeUrban Plan 172:47–59
10. Dai D, Rothenberg R, Luo R, Weaver SR, Stauber CE (2017) Improvement of geographic disparities: amelioration or displacement? J Urban Health 94(3):417–428
11. Tyler A (2015) Built environment and birth outcomes: examining the exposure to the Atlanta Beltline and its effects on community health. Thesis, Georgia State University School of Public Health
12. Fischer A (2018) Quantitative analysis of traffic related air pollution along the Atlanta Beltline east side trail. Thesis, Georgia State University School of Public Health
13. Pendergrast M (2017) City on the verge: Atlanta and the fight for America's Urban future. Basic Books, New York
14. Cardoni L (2013) The Atlanta Beltline: the beginning of an integrated bike network. Unpublished Master's thesis, Georgia Institute of Technology
15. Kanchik M (2017) Geographic information systems analysis of walkability data for the Atlanta Beltline communities
16. De Sousa CA (2003) Turning brownfields into green space in the City of Toronto. Landscape Urban Plann 62(4):181–198
17. Ghosh S, Byahut S, Masilela C (2019) Metropolitan regional scale smart city approaches in a Shrinking city in the American rust belt—case of Pittsburgh, Pennsylvania. In: Smart metropolitan regional development. Springer, Singapore, pp 979–1021
18. Pearsall H (2010) From brown to green? Assessing social vulnerability to environmental gentrification in New York city. Environ Plan C: Gov Policy 28(5):872–886
19. Higgins J (2008) Evaluating the Chicago brownfields initiative: the effects of city-initiated brownfield redevelopment on surrounding communities. Northwest J Law Soc Policy 3:240
20. De Sousa CA (2004) The greening of brownfields in American cities. J Environ Plann Manage 47(4):579–600
21. Raco M (2005) A step change or a step back? The thames gateway and the re-birth of the urban development corporations. Local Economy 20(2):141–153
22. Altherr W, Blumer D, Oldörp H, Nagel P (2007) How do stakeholders and legislation influence the allocation of green space on brownfield redevelopment projects? Five case studies from Switzerland, Germany and the UK. Bus Strategy Environ 16(7):512–522
23. Bae C-HC (2002) Orenco station, Portland, Oregon: a successful transit oriented development experiment? Transp Q 56(3):9–18
24. Atkinson R (2000) Measuring gentrification and displacement in greater London. Urban Stud 37(1):149–165
25. Cervero R, Duncan M (2002) Land value impacts of rail transit services in Los Angeles county. Report prepared for national association of realtors Urban land institute. https://drcog.org/documents/TODvalueLosangeles.pdf
26. Chatman DG, Tulach NK, Kim K (2012) Evaluating the economic impacts of light rail by measuring home appreciation a first look at New Jersey's river line. Urban Stud 49(3):467–487
27. Dawkins C, Moeckel R (2016) Transit-induced gentrification: who will stay, and who will go? Hous Policy Debate 26(4–5):1–18
28. Saunders P (2018) The scales of gentrification. Planning magazine
29. Immergluck D (2009) Large redevelopment initiatives, housing values and gentrification: the case of the Atlanta Beltline. Urban Stud 46(8):1723–1745

30. Auerbach J, Barton H, Blunt T, Chaganti V, Ghai B, Meng A, Blackburn C, Zegura E, Flores P (2017) Using data science as a community advocacy tool to promote equity in urban renewal programs: an analysis of Atlanta's anti-displacement tax fund. Available at https://arxiv.org/pdf/1710.02454.pdf
31. Immergluck D, Balan T (2018) Sustainable for whom? Green urban development, environmental gentrification, and the Atlanta Beltline. Urban Geogr 39(4):546–562
32. Weber S, Boley BB, Palardy N, Gaither CJ (2017) The impact of urban greenways on residential concerns: findings from the Atlanta Beltline trail. Landscape Urban Plann 167:147–156
33. Hsieh L, Noonan D (2018) The closer the better? Examining support for a large urban redevelopment project in Atlanta. J Urban Aff 40(2):246–260

# Ways and Means

# Exploring Livability in the Era of Smart City: A Case of Bhopal

**Amit Chatterjee, Gaurav Vaidya, N. K. Paulose, and Premjeet Das Gupta**

**Abstract** Bhopal is strategically located in Central India and also the Capital city of the State (Province) of Madhya Pradesh. Bhopal city is known for a city of knowledge, with culturally rich communities residing over generations, and also famous for many natural and artificial lakes. Out of the 100 cities chosen under the ambitious Smart Cities Mission by the Government of India, Bhopal has been selected in the first round as one of the 20 lighthouse cities. The Smart Cities Mission aims to transform the city into a future-ready smart city with enhanced livability and sustainability. The primary objective of this research is to assess the livability condition in Bhopal city based on following the three aspects: i) smart transportation, ii) influence of smart living on urban informality, and iii) good governance, including e-Governance and citizen participation. An attempt has also been made to assess livability score in the mobility sector and to understand smart practices with digital technology in the selected slum pockets and also to investigate the relationship between the socio-economic profile of administrative wards and the overall usage of digital governance apps in Bhopal city. The present research provides many clues for policymakers to understand livability conditions and user's perception towards promoting smart transport and governance, and that can call for a review of the smart city initiatives of Bhopal.

**Keywords** Smart cities mission · Livability · Mobility · Informality · Digital technology

A. Chatterjee (✉) · G. Vaidya · N. K. Paulose · P. Das Gupta
School of Planning and Architecture, Bhopal, India
e-mail: amit.chatterjee@spabhopal.ac.in

G. Vaidya
e-mail: gaurav.vaidya@spabhopal.ac.in

N. K. Paulose
e-mail: paulosenk@spabhopal.ac.in

P. Das Gupta
e-mail: premjeet@spabhopal.ac.in

 
T. M. Vinod Kumar (ed.), *Smart Living for Smart Cities*,
Advances in 21st Century Human Settlements,
https://doi.org/10.1007/978-981-15-4603-7_3

## 1 Introduction

Bhopal with an area of 463 km$^2$ and a population of 1.79 million is one of the fast-growing metropolitan cities in India. The city is divided into 85 wards. The name Bhopal was originated from Hindu King, namely Raja Bhoj, who governs the Malwa Region during 1018–1055. According to the popular myth, Raja Bhoj created a vast human made lake (Bhoj-tal or Bhoj's lake) covering 31 km$^2$, and Bhojtal became altered to Bhojpal and then Bhopal. The modern history of Bhopal originates with Sardar Dost Mohammad Khan, the founder of Bhopal state in 1715. The transformation from a village to a vibrant city-state started and continued till 1949 when the princely state of Bhopal was merged with an independent Indian Union. During more than 230 years of the Muslim dynasty, four women rulers (Begums or queens of Bhopal) govern Bhopal for almost 157 years with their strong personality, strong public support, and governing capabilities [15]. Being surrounded by the male-dominated world, part of conservative royal families and city dominated by mostly Hindu populations, the rise of female rulers and their success stories can be an excellent example of women's power in the glorious Indian history. In 1956, Bhopal was declared as the capital of Madhya Pradesh. During the journey of the city, from the nineteenth to twentieth Century, Bhopal transformed its identity from Nawabi Bhopal to the state capital of Madhya Pradesh, and then a city is known for its culturally rich communities, knowledge hubs and commercially active urban center with diverse economic population groups. This transformation was not possible without the contribution of all stakeholders, including urban poor in the city and their living and working conditions too.

Livability till a few years before was considered a realm of cities in the developed world. However, with the increasing awareness, technological innovation, and emphasis from all facades, especially the government, for sustainable development, the concept of livability has been gaining importance in the cities of the developing countries too [30]. Out of the 100 cities chosen under the Smart City Mission, Bhopal is one of the 20 cities to have been selected in the first round of smart cities challenge of the Government of India. The 'Mission Statement and Guidelines' of Government of India for smart cities states that: "In the imagination of any city dweller in India, the picture of a Smart City contains a wish list of infrastructure and services that describes his or her level of aspiration. To provide for the aspirations and needs of the citizens, urban planners ideally aim at developing the entire urban eco-system, which is represented by the four pillars of comprehensive development—institutional, physical, social and economic infrastructure. This can be a long term goal and cities can work towards developing such comprehensive infrastructure incrementally, adding on layers of smartness" [26, p. 5]. According to Smart city mission statement, the core infrastructure elements in a Smart City would include:

1. adequate water supply,
2. assured electricity supply,
3. sanitation, including solid waste management,
4. efficient urban mobility and public transport,

5. affordable housing, especially for the poor,
6. robust IT connectivity and digitalization,
7. good governance, especially e-Governance and citizen participation,
8. sustainable environment,
9. safety and security of citizens, particularly women, children and the elderly, and
10. health and education.

Starting 2015, the Smart City Mission envisions to transform the 'city of lakes', Bhopal into a future-ready Smart City with enhanced liveability, workability and sustainability. Government of Madhya Pradesh has incorporated a Special Purpose Vehicle (SPV), Bhopal Smart City Development Corporation Limited (BSCDCL) to plan, design, implement, coordinate and monitor the smart city projects in Bhopal. BSCDCL is a company incorporated under Indian Companies Act, 2013, with equal shareholding from Madhya Pradesh Urban Development Company Limited (MPUDCL) on behalf of Government of Madhya Pradesh (GoMP) and Bhopal Municipal Corporation (BMC). BSCDCL has received funds from Government of India and GoMP for the development of Bhopal as a Smart City.

The Area Based Development (ABD) proposals includes redevelopment and rehabilitation project for an 'already developed area'. The site chosen for the same is 145.8 ha of North and South TT Nagar. The project development is mainly focused along two major existing roads and a proposed 45 m Boulevard Street. The Boulevard Street will connect the two proposed commercial nodes at the metro stations. With an approximate investment of 2500 crores (462.96 million USD), the ABD area has been envisaged to develop with state-of-the-art infrastructure with all the smart features and green technology as per the smart city guidelines. These infrastructures will include 24 × 7 water supply and power, underground utility corridor, ICT infrastructure, smart street lighting, automated solid waste system, to name a few. Nearly 80% of the buildings in the area will be green rated. For efficient, advanced and commercially viable development plan in the 342 acres of Government land that emphasises on maximum utilization of space and optimum utilization of resources, Transit Oriented Development (TOD) guidelines, Urban and Regional Development Plans Formulation and Implementation (URDPFI) guidelines and Ministry of Housing and Urban Affairs (MoHUA) guidelines have been referred.

The citizen-centric innovations as a pan-city intervention is an 875 crores (162.04 million USD) project to collaborate, unify and ensure participation of all the citizens of Bhopal. An integrated command and control centre will monitor and manage the ABD area as well as the entire city. Bhopal Smart Plus Application and Smart Poles are meant to improve the information and communication technology. The app is a 90-layer GIS platform cutting across departments to keep in track property details, ensure public participation by adding city events, access ward-level information and send in suggestions and feedbacks. The 400 smart poles around the city act for Wi-Fi hotspots to carry out surveillance and provide lighting through LEDs. Heritage conservation, an initiative under PAN City Development includes restoration of Chawk Bazar around Jama Masjid Complex. The 5 m wide and 12 km long bicycle track along Hoshangabad road is integration between Bus rapid transit system (BRTS) and Public Bike Sharing System (PBS). The PBS Mobile application is an interface to check real-time availability and look for the nearest bicycle.

Bhopal is strategically placed at the center of India with excellent rail, road and air connectivity, an added advantage to attract investment and business. The Bhopal Smart City Mission envisions to Transform the 'city of lakes', Bhopal into a future-ready Smart City with enhanced livability, workability and sustainability. However, mobility alone cannot, but rather aspects such as planning for the informality, good governance, especially e-Governance and citizen participation determine the quality of life of the people and how liveable Bhopal is. The aim of this research is to access the livability condition based on following the three parameters: (i) Smart transportation; (ii) influence of smart living on urban informality; and (iii) good governance, especially e-Governance and citizen participation.

## 2 Demography

Bhopal city is the administrative headquarter of the Bhopal district. Bhopal city is the second most populous city (1.79 million) in the State of Madhya Pradesh, after Indore city (1.96 million) before extension as per the 2011 census. Recently, Bhopal Municipal Corporation (BMC) limit extended from 285 to 463 km$^2$, and the population increased from 1.79 to 1.88 million [14]. It is seen from Table 1 that since 1941, there has been a continuous growth of population in Bhopal city in absolute number till 2011. Bhopal grew at a high rate, and the population of Bhopal increased

**Table 1** Population growth and growth rate of Bhopal city (1941–2011)

| Year | Population (in million) | Decadal growth rate (in percentage) | Year | Population (in million) | Decadal growth rate (in percentage) |
|---|---|---|---|---|---|
| 1911 | 0.056 | – | 1981 | 0.671 | 74.35 |
| 1921 | 0.045 | −19.77 | 1991 | 1.063 | 58.51 |
| 1931 | 0.061 | 35.36 | 2001 | 1.423 | 33.84 |
| 1941 | 0.075 | 23.35 | 2011 | 1.798 | 26.31 |
| 1951 | 0.102 | 35.92 | 2020 (Projected)[a] | 2.361 | – |
| 1961 | 0.222 | 117.87 | 2025 (Projected)[a] | 2.650 | – |
| 1971 | 0.384 | 72.62 | 2030 (Projected)[a] | 2.959 | – |

*Source* CDP [7], Census of India [9]
[a]Projected by United Nations [40]

from 1.43 to 1.79 million during the decade 2001–2011. As per the UN-Habitat estimate, the projected population of Bhopal city will be 2.95 million by 2030 [40].

## 3 Transportation and Livability

Liveability is a significant component of the sustainability concept, which consists of six different objectives and components. One of the objectives is to achieve the transportation sector's goals, such as promoting public transportation, non-motorized transportation, giving more accessibility, and more transportation choice [41]. Hence, this shows that accessibility is a part of the liveability component in promoting a sustainable environment and creating a liveable place. Other than that, a liveable city promotes quality of access and linkages in neighbourhoods, town centre, and urban areas. The liveable city put emphasize emphasis on sustainability of transportation, which is to reduce noise and air pollution as well as encouraging residents to use low carbon mobility [17]. Land use transportation integration also plays a vital role in promoting smart urbanism.

Bhopal occupies a strategic location on the national transportation network which includes Rail, Road, Airways linking the city with various cities of the country as well as regional and district centers in the state [8]. The total number of new vehicle registrations in Bhopal RTO was 57,056 during the year 2013–14. There was an absolute growth of 10,732 (23.17%) in the new vehicle registrations over the previous year. Bhopal Road Transport Office accounted for 6.26% of total new vehicle registrations in MP during the year 2013–14. Out of the total registered vehicles, 86% of them comprised two-wheelers in 2002, but it has come down to around 73% in 2016 percentage. However, the private car ownership is fast increasing; out of the total registered vehicles in 2016, 17% of them were cars. A total number of registered buses in the city is 5193 in 2016 [25] (Fig. 1).

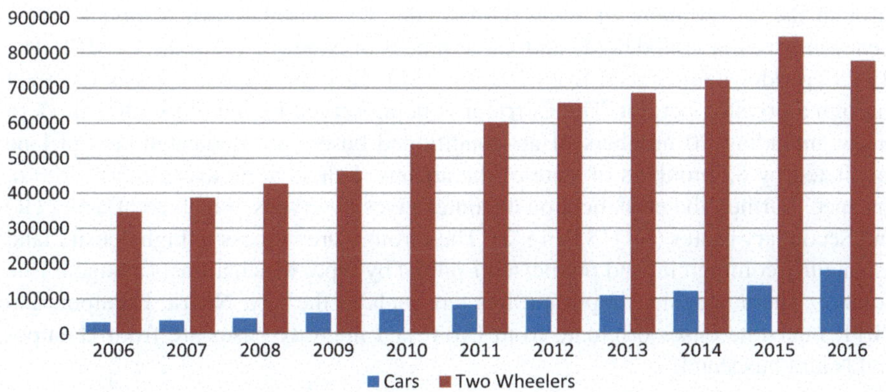


**Fig. 1** Number of registered cars and two wheelers in Bhopal. *Source* MoRTH [25]

**Table 2** Average trip length of various transport modes

| Trip length | | | | | | | |
|---|---|---|---|---|---|---|---|
| Mode | Walk | Cycle | Two-wheeler | Auto | Car/van | Mini bus | Standard bus |
| Trip length | 1.41 | 3.64 | 5.08 | 5.47 | 5.35 | 5.01 | 5.81 |

*Source* Comprehensive Mobility Plan [8]

The per capita trip rate (PCTR) is an indicator of the mobility of people. It is calculated in two ways, by including and excluding intra-traffic analysis zonal trips. PCTR, including the intra-traffic analysis zone in Bhopal is 1.37, and excluding intra-zonal trips is 0.96 [8]. The estimated trip generated in Bhopal is 2.46 million, considering the population of 1.8 million population of the city. Average trip lengths of various modes of transport are given in Table 2. The average trip length of buses and mini-buses are around 6 km.

The distribution of trips by purpose and mode is shown in Table 3. About 50% of the trips are for work purposes while education trips account for 28% of the journeys. Two-wheelers are used for most of the work trips (20.70%). Around 43 percentages of the trips are made by walk followed by 23% of the trips are made by public transport and intermediate public transport.

### *3.1 Public Transport and Intermediate Public Transport*

Public Transport modes available in Bhopal included BRTS, Standard Buses, Intermediate Public Transport (Mini-Buses), and Para Transit (Vans and Shared Autos). The total length of the BRTS corridor is 28 km which starts from Bairagarh to Misroad. All other arterial and sub arterial roads are having normal buses. Including BRTS and the Bus services provided by BRTS are only available at eight routes, which are operated through 250 busses as on by August 2018. The project is financed through the Government of India (GoI) under Jawaharlal Nehru National Urban Renewal Mission (JNNURM) and Government of Madhya Pradesh (GoMP). The BRTS corridor commenced from 2013 as "MY Bus" on Net-Cost Basis Contract through a private operator. The Corridor is being served by low floor ultra-modern buses, including 20 numbers of air-conditioned buses in a dedicated central lane facilitated by 82 numbers of state of the art bus shelters at an average 500–700 m distance. Further, the classification of routes gives two types, which are Trunk (TR) and Secondary routes (SR) [8] (Fig. 2). These routes are witness of high density land use mainly commercial and public/semi-public by type, which attracts a huge traffic volume. There are three important bus terminals in the city, Nadra, Halalpur, and ISBT. Interstate buses originate from ISBT, and intercity buses are from Halalpur and Nadra bus stand.

**Table 3** Distribution of trips by purpose and modes

| | Work | Business | Education | Social | Shopping | Recreation | Health | Other | Total |
|---|---|---|---|---|---|---|---|---|---|
| Walk | 13.80 | 0.50 | 20.20 | 0.90 | 7.2 | 0.00 | 0.10 | 0.30 | 43.00 |
| Cycle | 2.20 | 0.10 | 1.20 | 0.00 | 0.30 | 0.00 | 0.10 | 0.00 | 4.00 |
| Two-wheeler | 20.70 | 1.00 | 1.20 | 0.50 | 1.10 | 0.10 | 0.20 | 0.40 | 25.00 |
| Auto | 0.70 | 0.30 | 0.00 | 0.00 | 0.20 | 0.00 | 0.10 | 0.10 | 1.00 |
| Car | 1.60 | 0.20 | 1.30 | 0.00 | 0.20 | 0.00 | 0.10 | 0.10 | 3.00 |
| Mini bus | 8.90 | 0.00 | 2.80 | 1.20 | 5.60 | 0.00 | 1.10 | 0.60 | 20.00 |
| Standard bus | 1.90 | 0.1 | 0.80 | 0.00 | 0.40 | 0.00 | 0.00 | 0.00 | 3.00 |
| Overall | 49.80 | 2.10 | 27.50 | 2.60 | 14.90 | 0.10 | 1.60 | 1.40 | 100 |

*Source* Comprehensive Mobility Plan [8]

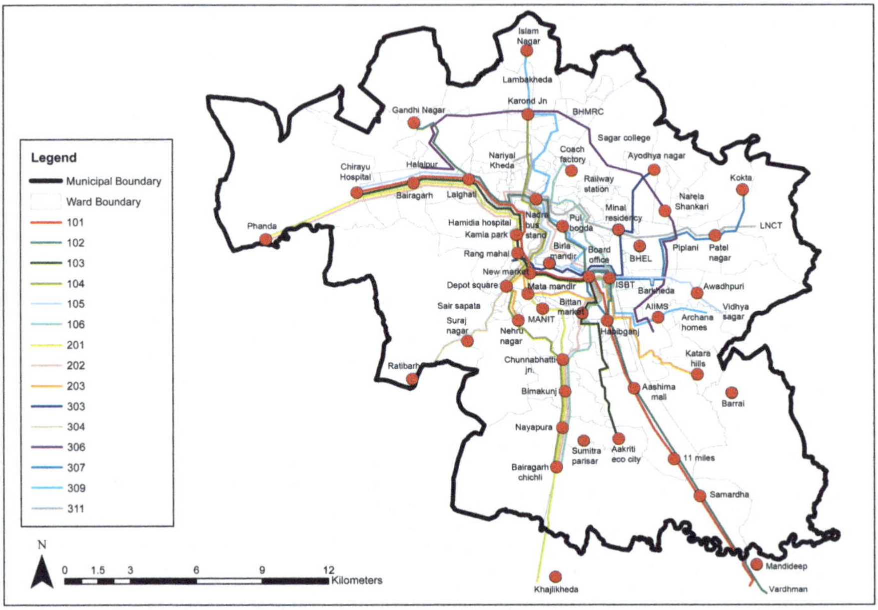


**Fig. 2** Public transport routes in Bhopal. *Source* BCLL [2]

Bhopal City Link Limited (BCLL) is a Special Purpose Vehicle (SPV) owned by Bhopal Municipal Corporation (BMC) was incorporated in 2006 to act as the nodal agency to improve public transportation in Bhopal. The Present Operational Bus Services by Bhopal City Link Limited headed under the chairpersonship of Mayor of Bhopal, Commissioner BMC as Managing Director and dedicated Chief executive officer of the company. BCLL has been fulfilling the municipal responsibility of providing Public Transport facility for the conveyance of public mandated under Sec 66 (S) of Municipal Corporation Act. BCLL's 'My Bus' is equipped with state of the art GPS system, Passenger information system, Passenger announcement system, CCTV cameras. Bus priority signaling system and pedestrian crossing signal system is provided. Dedicated Squads for Safety and Efficiency of Bus Operation has been deployed. All low floor buses are tracked in the control room on a real time basis. The vehicle tracking unit devices have been installed on all buses. GPS receiver modules, for bus stop announcement in the buses, have also been installed on all the buses. The focus of planned Intelligent Transport System (PITS) is to automate the BCLL's My Bus operations utilizing a GPS based vehicle Tracking and GIS-based monitoring System and thereby to improve the frequency and the service level benchmarks of the system.

The Automated Fare Collection and Surveillance System for buses and bus stops is a part of the Intelligent Transport System established in Bhopal. There are 11 Point of the Sales counter is established in different parts of cities for issuance of Smart Pass for commuters of Bhopal. Automated fare gates (AFG) and Automatic Ticket

**Table 4** Rate of subsidy for monthly passes in Bhopal

| S. No. | Category | Subsidy (%) | Amount |
|---|---|---|---|
| 1 | General public | – | Rs. 800 |
| 2 | Senior citizen | 50 | Rs. 400 |
| 3 | State govt. employees | 50 | Rs. 400 |
| 4 | Press reporters | 50 | Rs. 400 |
| 5 | Women | 50 | Rs. 400 |
| 6 | Students (age less than 35 years) | 62.50 | Rs. 300 |
| 7 | Handicapped (more than 25%) | 75 | Rs. 200 |
| 8 | Bhopal municipal corporation currently working employees | 75 | Rs. 200 |
| 9 | Beneficiaries in govt. schemes | 50 | Rs. 400 |
| 10 | Labour cardholder | 50 | Rs. 400 |
| 11 | Prepaid card | – | Rs. 200 (minimum) |

*Source* BCLL [2]
*Note* One USD is INR 70

vending Machines (ATVM) are established in terminals in the bus stops. To ease the travel journey in Bhopal, Bhopal city bus started options like Bhopal City Bus Monthly Pass Named Mahapaur Smart Pass and BRTS App. According to MY Bus web, there are total of 19,601 Active Cards Users. The price of the Bhopal City Bus is INR 800/- for monthly smart pass. This pass is valid for all AC and Non AC city buses of Bhopal Municipal Corporation for one month from the date of the card issued or recharged. For Women, state govt. employees, senior citizen and labor cardholder My Bus smart pass is available with a 50% subsidy. Students get a 62.50% subsidy in the monthly pass (Table 4).

In addition to Trunk and Standard routes, 11 mini buses routes are also operational in Bhopal (Fig. 3). Around 507 mini buses are providing mobility service in these routes. Shared Autos and vans are also helping to meet the derived demand for transport. There are 15 major arterial and sub-arterial roads in which this service is available.

Bhopal Metro rail project will include a total of six lines in a phased development. Phase one has already been approved. The phase one will involve two corridors with a combined length of 27.87 km (Table 5). The first corridor will be approximately 15 km-long and connect Karond Circle to AIIMS (Line 2 or Purple Line). The second 12.88 km-long corridor will connect Bhadbhada Square to Ratnagiri Tiraha (Line 5 or Red Line). Line 2 will be mostly elevated apart from the Bhopal railway station and bus station, where it will be partially underground. The second corridor will feature a total of 16 stations, including 14 elevated and two underground.

Other Proposed lines include Bairagarh to Awadhpuri (Line 1), Airport to Vasant Kunj (Line 3A), and Bhauri Bypass to Vasant Kunj (Line 3B). Line 4 will extend from Ashoka Garden to Mother Theresa School. Line 6 will connect Mandideep to Habibganj Station. This system consists of a network of 105 km (Fig. 4). A common central

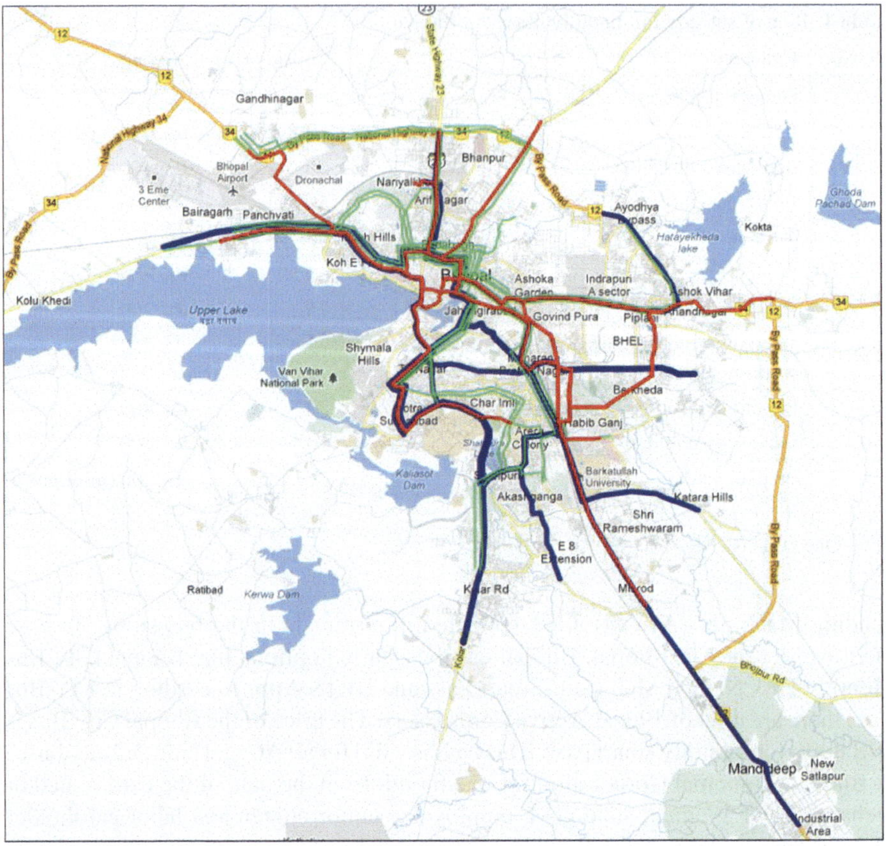


**Fig. 3** Intermediate public transport routes in Bhopal. *Source* Comprehensive Mobility Plan [8]

**Table 5** Bhopal metro phase-I

| Route name | Route description | Stations in number | | Length (km) | |
|---|---|---|---|---|---|
| Purple line | Karond Square to AIIMS | Total | 16 | Total | 14.99 |
| | | Elevated | 14 | Elevated | 12.58 |
| | | Underground | 2 | Ramp | 0.62 |
| | | | | Underground | 1.79 |
| Red line | Bhadbhada Square to Ratnagiri Tiraha | Total | 14 | Total | 12.88 |
| | | Elevated | 14 | Elevated | 12.88 |
| Total | | | 30 | | 27.87 |

*Source* MPMRCL [20]

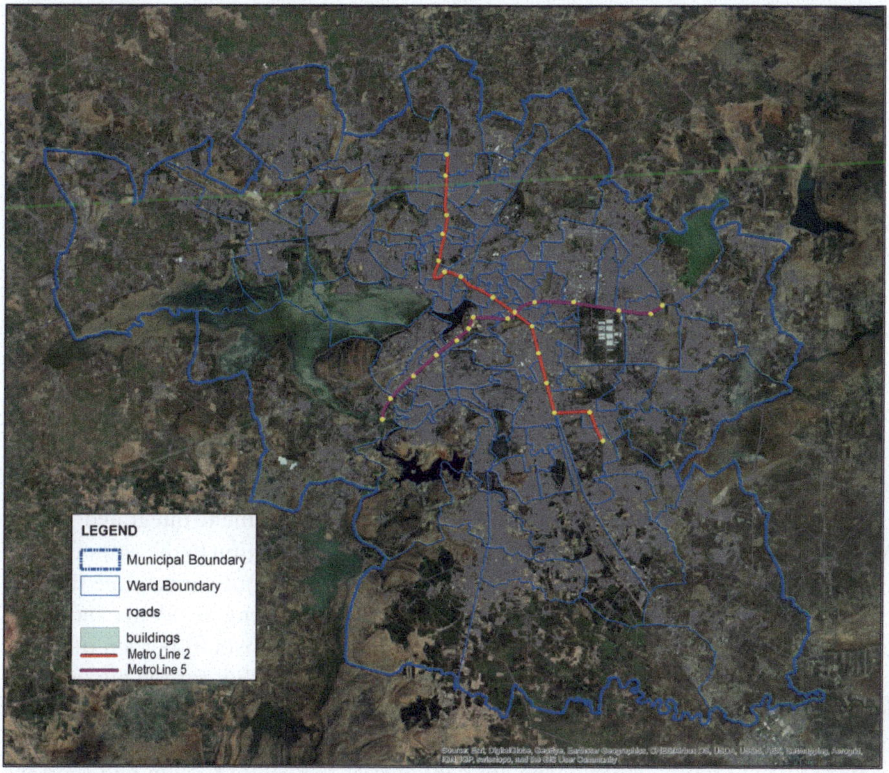


**Fig. 4** Proposed metro routes, Bhopal (Phase I). *Source* MPMRCL [20]

depot will be built near Subhash Nagar Underpass for the Purple and Red Lines. The depot will have facilities for stabling, inspection, and vehicle maintenance. Bhopal Metro corridors will be integrated with railway stations and bus stations, as well as a feeder network of buses, intermediate public transport (IPT), and non-motorised transport (NMT) to improve accessibility. The Central and the Madhya Pradesh governments provided funding for the project on an equal equity basis. MPMRCL also proposes to secure loan from the European Investment Bank (EIB) for the project. In addition to revenue from ticket fares, the project will also receive non-fare box revenue through rentals and advertisements. Other revenue sources include value capture financing (VCF) through a mechanism of transit-oriented development (TOD) and transfer of development rights (TDR). Estimated daily ridership in Purple Line is 0.21 million in 2021, and in the Red line the daily ridership would be 0.6 million (Table 6).

**Table 6** Projected ridership, Bhopal metro project

| Purple line | | | | Red line | | |
|---|---|---|---|---|---|---|
| Year | Peak hour ridership | Number of passengers daily | Passenger (km) | Peak hour ridership | Number of Passengers daily | Passenger (km) |
| 2021 | 21,522 | 215,224 | 968,194 | 13,885 | 138,848 | 646,481 |
| 2031 | 28,370 | 283,703 | 1,276,250 | 18,303 | 183,026 | 852,176 |
| 2041 | 37,397 | 373,970 | 1,682,322 | 24,126 | 241,260 | 1,123,318 |
| 2054 | 53,553 | 535,525 | 2,409,085 | 34,548 | 345,484 | 1,608,591 |

*Source* MPMRCL [20]

## 3.2 Smart Road and Traffic Management System

As part of smart city initiatives Bhopal Smart City Development Corporation Limited (BSCDCL) has taken up project for Smart traffic management-Integrated Traffic Management System (ITMS) which provide greater information to the authorities to proactively manage the ongoing traffic situation, enhance traffic discipline through effective enforcements and increase road safety by preventing accidents and allow citizens to make informed travel choices. With ITMS, the overall burden of traffic on the roads will reduce, and will result in enhanced reliability for the citizens and local businesses, improving overall productivity levels [6]. The ITMS command and control room is set up at the Bhopal Smart City Development Corporation (BSCDCL) centre at Govindpura. Traffic cops are deployed at the centre to sign and send the challans. Since its start on June 13, 2018, more than 0.18 million e-challans have been issued until August 15, this year [22]. The plan is to install 240 smart cameras in 22 squares [6].

The Project of Smart road initiated by BSCDCL keeping in mind to serve the peoples for road connectivity along with smart features like wise smart poles to serve internet facilities and lighting with CCTV vigilance, weather reporting and pollution level detector. These poles have been installed 5 in numbers at equidistance starting from Bharat Mata square to ploy technique square. This Smart Road will serve as a cycle track keep in mind to used non-motorized vehicle uses having width of cycle track 2.5 m on both sides of the carriage-way. Another important feature of smart road is underground electrification for LT lines inside duct having 1.2 m depth and 1 m width over both sides of main carriage way throughout the length of road. There are 2 lanes separate 8.5 m main carriage way with DBM BC crust having a camber of 2.5% which will serve as smooth surface runoff of rain water from road surface. This crust has been designed for the speed of 80 kmph. There is also a footpath provision for pedestrian's walkway having 2.0 m effective width on both sides throughout the length of the road [6].

## 3.3 Public Bicycle Sharing

Bhopal Smart Public Bike Sharing is India's first fully automated system. The stations are crewless and linked to the Central control system. This data is used to make decisions on the redistribution of cycles around stations during the hours of operations. It is fully working with online payment. There are 500 cycles and 75 stations. It is physically integrated with BRTS Service in the city.

The fee for using one hour is INR 20.00 and it can be paid by online banking services (Table 7). The data analysis of the users shows that around 60% of the trips are less than 30 min. Close to 20% of the trips are between 30 min to one hour [6] (Fig. 5). The public bicycle hubs in Bhopal are shown in Fig. 6 and the proposed plan for bicycle sharing facilities augmentation is shown in Table 8.

**Table 7** Bicycle usage fare

| S. No | Time | Non-member (Rs.) | Member (Rs.) |
|---|---|---|---|
| 1 | <30 min | 10 | 0 |
| 2 | Till 60 min | 20 | 10 |
| 3 | Till 90 min | 40 | 20 |
| 4 | Till 120 min | 60 | 30 |
| 5 | Every additional 30 min | 30 | 15 |

*Source* Smart City Cell, Bhopal
*Note* 1 USD = INR 70

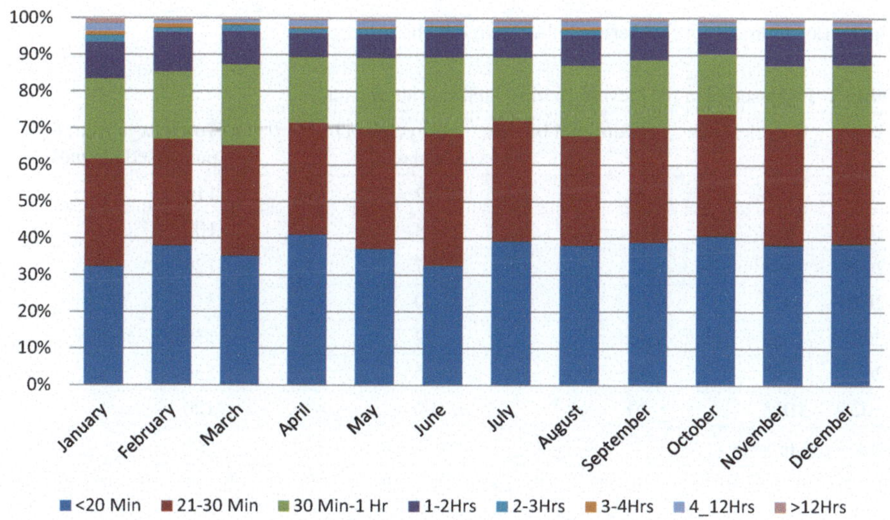


**Fig. 5** Month-wise duration of bicycle usage in Bhopal. *Source* BSCDCL [6]

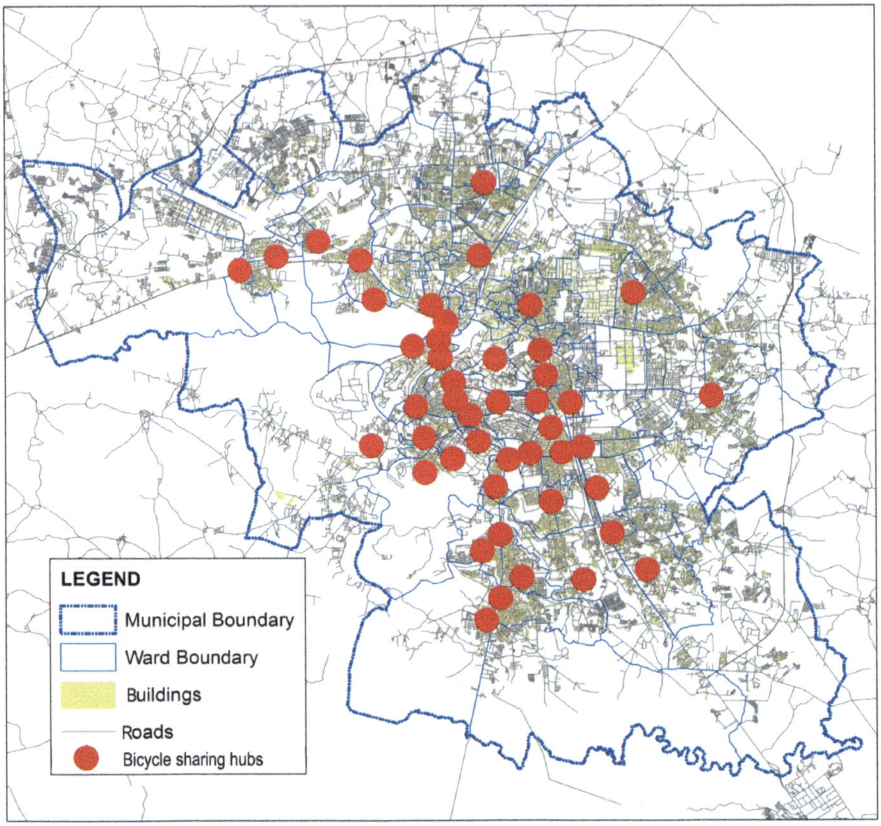


**Fig. 6** Location of bicycle sharing hubs. *Source* BMC [5]

**Table 8** Proposed plan for bicycle sharing facilities augmentation

| Year | Bicycles/year | Additional bicycle | Bicycles to be replaced | Total bicycles to be purchased annually |
|---|---|---|---|---|
| 2017 | 850 | 30 | 82 | 110 |
| 2018 | 865 | 15 | 85 | 100 |
| 2019 | 900 | 35 | 87 | 120 |
| 2020 | 920 | 20 | 90 | 110 |
| 2021 | 960 | 40 | 92 | 130 |
| 2022 | 995 | 35 | 96 | 130 |
| 2023 | 1015 | 20 | 100 | 120 |

*Source* BMC [5]

## 3.4 Smart Parking Management

According to Bhopal Municipal Corporation (BMC) report, At Present, about 9184 vehicles are parked on streets, while 4548 are parked off-street, and about 28,302 vehicles are parked illegally. Private parking facilities cater to less than 10,000 vehicles in Bhopal. BMC is in the process of implementing smart parking solutions in the city. Smart Parking project would create an impact on the ease of mobility of people with integrated smart parking solutions, which enables quick, faster accessibility with single space detection, parking guidance on real-time and allow the commuters to reserve the parking slot. This would also provide a smart solution which would reduce the travel time, carbon emission, search time, traffic congestion in the city. Besides, it also improves traffic flow within the parking lot with efficient design and creating. Locating a parking space with clear and simple direction and variable message reduces traffic congestion. Quick entry and exit to the parking lots (on street, off street and multi-level parking) with lesser transactional time [39]. Bhopal smart parking system provides information about the availability of parking slots on real-time with the help of a web portal/parking App; around 60 locations across the city would be managed through public-private partnership mode under the smart city parking. Close to 20 parking locations have started functioning under smart parking city parking solutions. There are 10 other locations identified for Multi-Level Parking Plaza [6] (Fig. 7; Table 9).

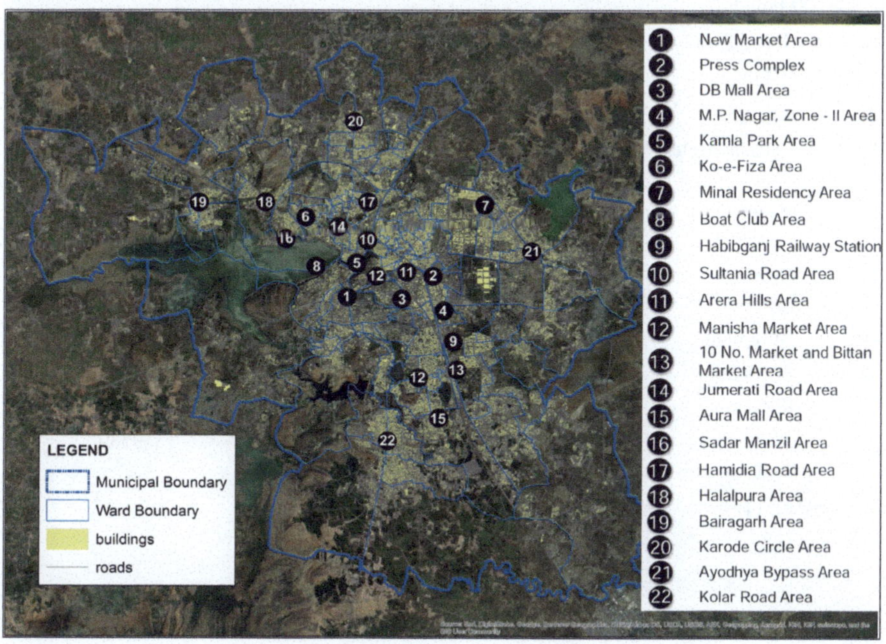


**Fig. 7** Smart parking locations. *Source* BMC [4]

**Table 9** Major parking demand locations

| S. No. | Area | Legal off-street | Legal on street | Illegal on street | Private parking |
|---|---|---|---|---|---|
| 1 | New market area | 863 | 869 | 2038 | 1092 |
| 2 | Boat club | | 165 | | 65 |
| 3 | Manisha market | 407 | 145 | 1181 | 160 |
| 4 | Bittan and 10 no. market | 845 | 614 | 1350 | 444 |
| 5 | HBJ railway station | 1151 | | 980 | 182 |
| 6 | MP nagar zone 2 | 1103 | 459 | 2901 | 64 |
| 7 | DB mall area | 1040 | 501 | 1449 | 844 |
| 8 | Press complex | 353 | 288 | 1540 | 492 |
| 9 | Kamla park | 180 | 10 | 1645 | 60 |
| 10 | Sultania road | 251 | 19 | 2063 | |
| 11 | Jumerati road | 233 | 16 | 1546 | |
| 12 | Hamidia road | 414 | 789 | 1504 | 285 |
| 13 | Royal market and Sadar Manzil | 601 | | 2442 | 466 |
| 14 | Koh-e-Fiza | 7 | 37 | 367 | 390 |
| 15 | Halalpura | 334 | | 56 | 597 |
| 16 | Bairagarh | 222 | | 1086 | 281 |
| 17 | Karodh circle area | | | 3161 | |
| 18 | Minal residency | 293 | 93 | 94 | 70 |
| 19 | Ayodhya bypass | 154 | | 302 | 19 |
| 20 | Aura mall | 101 | 70 | 676 | 960 |
| 21 | Arera hills | | | 991 | 2657 |
| 22 | Kolar road | 400 | 75 | 735 | 407 |
| | All inventor areas | 9184 | 4548 | 28,302 | 9539 |

*Source* BMC [4]

### 3.5 Bhopal Livability Score: Mobility

Ministry of Housing and Urban Development has come up with a total of 79 Indicators (57 Core Indicators and 22 Supporting Indicators) for measuring the livability of cities. There are Core Indicators that are considered an essential measure of the livability of cities, and the Supporting Indicators to supplement the Core Indicators by adding value to them. There are 12 indicators in the mobility sector. An assessment of Bhopal shows that the city is slowly progressing towards better livability. Bhopal has only 15% of its geographical area covered under the public transport system, and the availability of public transport buses per 1000 people is also low. Bhopal scores relatively better in terms of modal share of non-motorized transport trips. The modal share of public transport and intermediate public transport trips is 23% (Table 10).

From 2004 to 2009, accidents have continuously increased, and road accidents in Bhopal have peaked in the year 2009 at 3719. Deaths due to road accidents were highest in 2012. Total accidents, deaths, and the number of injured people have decreased. In 2017 the total number of reported accidents had come down to 3393 and deaths to 242. The decrease in road accidents loss life and injured people can be connected to the travel demand management measures adopted in the past one decade in terms of improvement in public transport, creation of pedestrian and bicycling infrastructure, and introduction of the smart traffic management systems.

**Table 10** Livability score, Bhopal

| S. No. | Transportation and mobility | Score |
|---|---|---|
| 1 | Geographical coverage of public transport % | 15.00 |
| 2 | Availability of public transport buses per 1000 persons) (supporting) (including the mini buses) | 3.94 |
| 3 | Mode share of public transport (core) in % (including the mini buses) | 23.00 |
| 4 | Percentage of road network with dedicated bicycle tracks (core) | 1.00 |
| 5 | Percentage of interchanges with bicycle parking facilities (supporting) | Nil |
| 6 | Mode share of non-motorized transport (core) | 47.00 |
| 7 | Availability of passenger information system (supporting) | 10.00 |
| 8 | Extent of signal synchronization (supporting) | Nil |
| 9 | Availability of paid parking spaces (core) | 45.12 |
| 10 | Percentage coverage of footpaths—wider than 1.2 m (core) | 20.00 |
| 11 | Percentage of traffic intersections with pedestrian crossing facilities (supporting) | 15.00 |
| 12 | Extent to which universal accessibility is incorporated in public rights-of-way (supporting) | 10.00 |

*Source* Calculated by the Authors

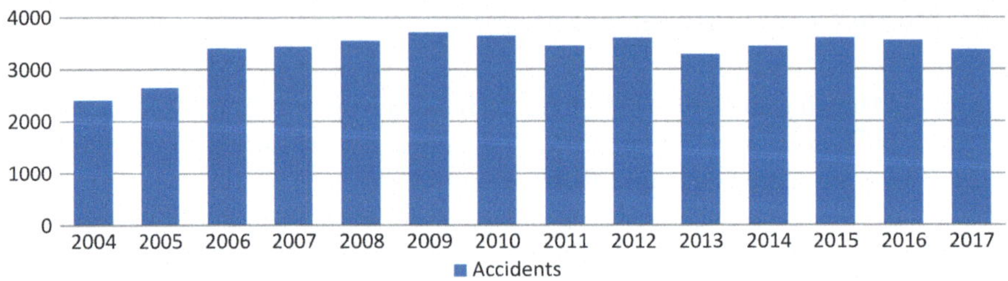


**Fig. 8** Road accident trend in Bhopal (2004–2017). *Source* MoRTH [25]

In that way, it has given inducement to improvement in the livability in Bhopal (Fig. 8; Table 11).

Measures adopted by BMC to improve the public transport system in the city have benefited many commuters. Close to 0.4 million trips are undertaken by public transport and intermediate public transport. Around 90% of the economically weaker section household and informal settlements are in the influence area of the Bhopal Mybus corridor (Fig. 9).

## 4 Influence of Smart Living on the Urban Informality

One can understand and visualize urban as an area of high population density, with a large scale superior infrastructure and services, offering better employment opportunities without predominance of agricultural activities. But at the same time, inherent economically weaker groups of urban areas and continuous migration from rural areas in search of employment are creating a different set of identity. These people with limited or less resources are living in informal urban conditions, sometimes on the outskirts of cities, and work in the informal employment sectors without any financial and social security. A habitation cannot attain its best progress on the various dimensions of development without addressing the needs of all stakeholders from different socio-economic groups. Therefore, it is important to consider the concerns of economical weaker groups of society whenever inclusive growth is desired [21]. Planning for Urban Informality is one of the approaches adopted in recent years, which covers 'Unplannable' and challenging aspects of human habitation like; informal housing and informal employment for vulnerable groups who are dependent on the same and have their 'Right to the city' as similar to any other citizen of that urban settlement [33].

As per the statistics, around 22% of people live in urban areas on incomes that are below the poverty line. At the national level, rural poverty remains higher than urban poverty, but the gap is closing [19]. Global Multi-Dimensional Poverty Index 2018 has ranked Madhya Pradesh as the fourth poorest state in the country. According to the erstwhile Planning Commission of India, more than 42% of the urban population of Madhya Pradesh state is designated Below Poverty Line (BPL), compared to 37%

**Table 11** Reported accidents in Bhopal

| | 2004 | 2005 | 2006 | 2007 | 2008 | 2009 | 2010 | 2011 | 2012 | 2013 | 2014 | 2015 | 2016 | 2017 |
|---|---|---|---|---|---|---|---|---|---|---|---|---|---|---|
| Accidents | 2412 | 2651 | 3409 | 3448 | 3554 | 3719 | 3653 | 3459 | 3623 | 3294 | 3459 | 3621 | 3571 | 3393 |
| Death | 171 | 208 | 184 | 254 | 279 | 272 | 270 | 299 | 301 | 257 | 236 | 235 | 237 | 242 |
| Injured | 2537 | 2571 | 2537 | 3163 | 3305 | 2990 | 2990 | 2990 | 3076 | 3076 | 2651 | 2779 | 2481 | 2465 |

*Source* MoRTH [25]

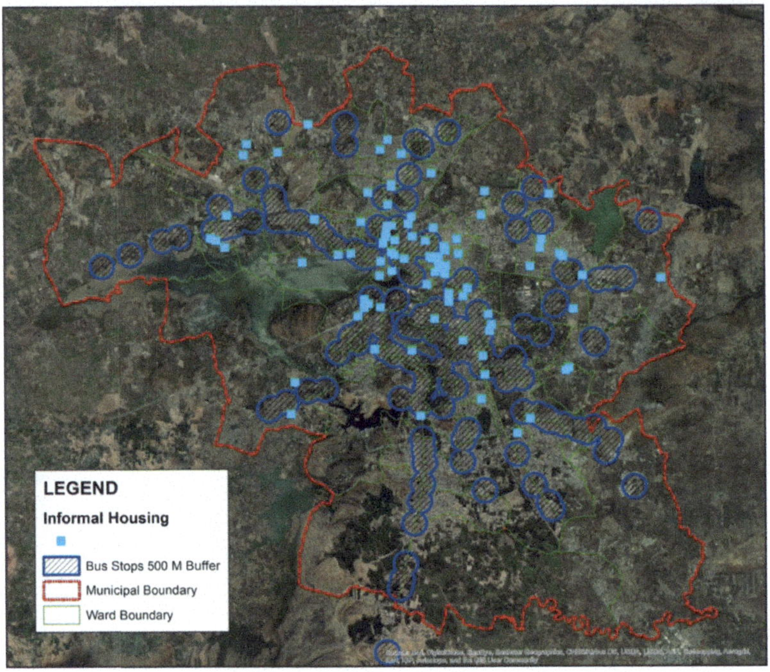


**Fig. 9** Public transport bus stops and informal housing locations. *Source* BMC [3] and BCLL [2]

in rural areas of the state. Urban poor in the capital city Bhopal constitutes 25% of the total population of the city out of which 15% are below poverty line as per definitions of poverty line by C. Rangarajan committee [1]. Urban informality is typically ascribed to the urban poor in the cities belonging to dispossessed 'marginalized' working class surviving with informal housing and living conditions [24].

Housing is considered as informal when the resources that the households command within the house are insufficient, to enable them to consume sufficient goods and services for reasonable minimum level of welfare. Housing poverty is defined by guidelines of Rajiv Awas Yojana (RAY) scheme as; Individuals and households who lack safe, secure and healthy living environment with basic infrastructure such as piped water and adequate provision of sewerage sanitation, drainage and removal of household waste. Informal sector housing is witnessed mostly in the shape of jhuggi/jhopdi settlements, in general also known as slums and squatters [16].

The informal sector is described as, 'Unincorporated or unregistered employment activities, operated without any labour social and financial protection through their work' [10]. A report of the National Commission for Enterprises in the Unorganised Sector by the Government of India [36] finds that about 86% of the total employment in 2004–2005 was in the informal sector, whereas a new report of International Labour Organisation [13] claims that close to 81% of all employed persons in India are working in the informal sector. The National Council of Applied Economic

Research of India estimates that the informal sector—"unorganised sector"—generates about 62% of GDP, 50% of national savings and 40% of national exports. In terms of employment in India, the informal economy provides total employment of about 75% in rural areas and 69% in urban area as per the estimates of 68th round of periodic employment and unemployment surveys (EUS) conducted by National Sample Survey Organisation in year 2011 12 [27]. Therefore, informal employment has rapidly become an important topic of consideration over the last few decades. The National Statistical Commission (NSC) states that the reason why a huge proportion of the poor in India is getting engaged in the informal sector is because of lack of basic level of education and skill, difficulty to get a job in a formal sector which leads to large-scale migration and the sprouting of slums in several Indian cities. Another factor is that the growth rate of jobs being created in less than the rate at which these are growing. It is also seen that sometimes an informal worker is as qualified as a formal employee but due to lack of opportunities, his work is not getting recognized under the tag of formal sector [19].

World is changing every day in age of globalization and new approaches of urban development are adopted all over with latest technology to support modern life styles of mankind. One of the recent idea is named as "Smart Living", which explains how technology can be helpful in achieving growth and, convenience and comfort with all means of quality of life, i.e.; access to good health, education and housing quality, physical infrastructure and services, individual safety and security, connected communities and access to communication and system awareness by innovation and experimenting for sustainable living and smart care. Post millennium with effort of government, communities and system as a whole, quality of life and exposure to modern world is changing for urban poor. Now this urban informality is also transforming its character and getting influenced by the idea of smart living.

In last two decades, a lot of efforts were made by the central and the state governments with active participation of communities for the upliftment of urban poor and urban informality as a whole. Present scenario of access to information, communication, awareness and technological reach of common citizens are contributing for growth oriented development in lives of people for all sections of social strata [31]. With the changing world, people living in informal housing and working in informal employments are also getting influenced by global influence of smart living through economic, technological and social transformation of cities. Depiction of the impact of smart living on urban informality is illustrated through example of Bhopal as a case study with focusing on the role of communication and information technology and latest financial transaction practices by the people of this socio-economic group.

Influence of smart living on urban informality is demonstrated with detailed assessment of two key sectors, i.e. Informal Housing and Informal Employment. Therefore, a field investigation was undertaken, of four character-wise varied informal housing settlements and four market areas—each one is unique in nature for informal trades and working population within Bhopal municipal limits. Bhopal has witnessed transformation in its identity from being Nawabi Bhopal to the state capital of Madhya Pradesh and then a city known for its culturally rich communities, knowledge hubs and commercially active urban center with diverse economic population groups from nineteenth to twentieth Century. This transformation was not possible without the contribution of all stakeholders, henceforth population at the bottom of the pyramid is equally important and so are their living and working conditions.

## *4.1 Informal Housing*

As per Census 2011, 26.68% population of Bhopal city, with more than one lakh households (102,803 HHs) is living in informal housing, some of them occupying strategic locations of government owned lands and critical drainage basin along the lakes shores, river banks and nallahs. There are 384 notified and non-notified slums in Bhopal city [7], out of which 36% of slum population is squatting in the new Bhopal area mostly on government lands, whereas around 33% of the slum population lies in the Old Bhopal area and remaining 30% of the population is sparsely distributed in the fringes and Outgrowths [34]. The location aspect of the slum settlements reveals that the slums lie in the strategically located and well-developed areas close to the work centers. The slums in Bhopal are having low to medium density. In almost every slum, the peripheral area of the slum was found to be very dense whereas in the interiors, there were open spaces.

To understand the impact of smart practices on urban poor as well as urban housing as a whole, four different informal settlements were selected for the field investigation purpose, named are; Roshanpura slum, Rahul Nagar slum, Gondipura slum and Kohefiza slum. Stratified proportionate random sampling method is adopted for primary socio-economic surveys of 240 households (60 samples each from the four selected slums). At the same time, built-up and service inventory of each slum is prepared to understand the slum character (Fig. 10).

Roshanpura slum is a notified slum, located in ward no. 24 in the central part of the city near New Market. This slum is spread over 6.32 ha area with around 2300 households and more than 11,000 slum dwellers. Roshanpura slum came in existence in 1961, during the development of new capital city complex and is known as one of oldest slums in Bhopal. The elevation varies from 512 to 540 m in this slum area and this large difference between the highest and lowest point have impact on housing quality and socio-economic condition of households. This slum evolved on government land, reserved for public and semi-public development purpose in Bhopal master plan 2005, which is still under enforcement status. Majority of families are living in the slum from more than 40–50 years and more than 80% dwellings were in pucca and semi-pucca condition (Fig. 11).

'Rahul Nagar slum' is also a notified slum, located in ward no. 29 of city near Maulana Azad National Institute of Technology, Bhopal. This slum is spread over 8.13 ha area with around 142 households and more than 700 residents. Rahul Nagar slum came in existence in 1970s and hence, considered as another of oldest slums of Bhopal. The elevation varies from 507 to 524 m in this slum and this large difference between the highest and lowest point have an impact on housing quality and socio-economic condition of households. This slum evolved on government land, reserved

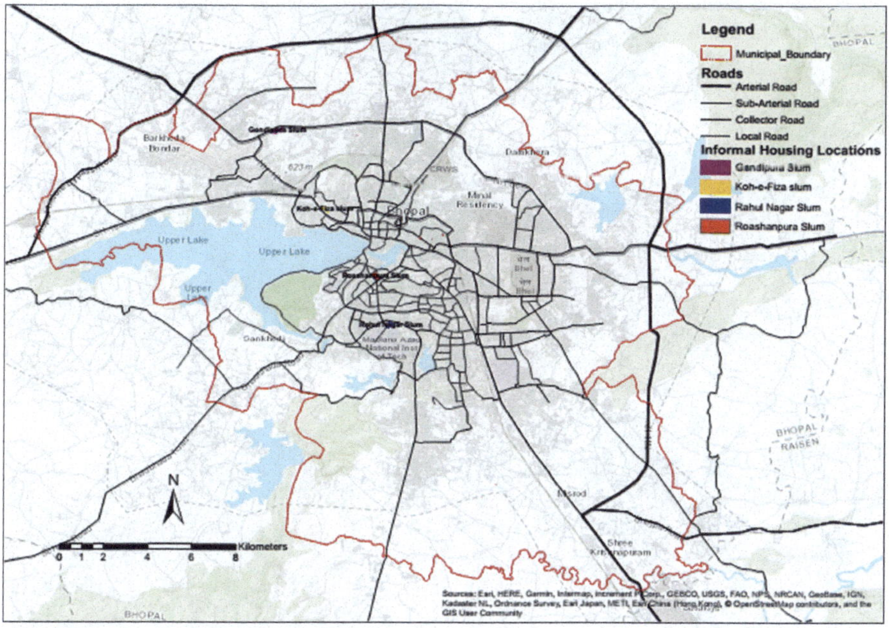


**Fig. 10** Locations map of the selected slums for the study of informal housing in Bhopal *Source* School of Planning and Architecture [35]

for residential and commercial development purpose in Bhopal master plan. Majority of families are living in the slum from more than 40–50 years and more than 70% dwellings were in pucca and semi-pucca condition, partial (35–40%) slum dwellings were rehabilitated in the site under Rajiv Awas Yojana in year 2010 under in situ development (Fig. 12).

'Gondipura slum' is also a notified slum, located in ward no. 75 at northern part of city near Rajiv Gandhi Prodgyogiki Vishwavidyalaya, Bhopal. This slum is spread over 9 ha area with around 180 households and more than 800 residents. Gondipura slum came in existence in 1990s, initially evolved by nomads of Bheel tribes on slopes and later slum grew as the people from Ratibad were shifted here. Western part of the slum with land of 1.5 ha is well distributed in plotted development under slum rehabilitation project for Bheel tribes and remaining slum has evolved organically. The elevation varies from 515 to 550 m in the slum and this large difference between the highest and lowest point have an impact on housing quality and socio-economic condition of households. This slum evolved on government land, kept reserved for, mainly as green-belt (non-developable due to steep slopes) and mix-landuse along the periphery in Bhopal master plan 2005. Majority of families are living in the slum from more than 25–30 years and more than 60% dwellings were in pucca and semi-pucca condition (Fig. 13).

'Kohefiza slum or Idgah Hills Slum' is a non-notified slum located in ward no. 10, evolved along the Sultaniya road ahead of Lalghati square towards Bhopal district

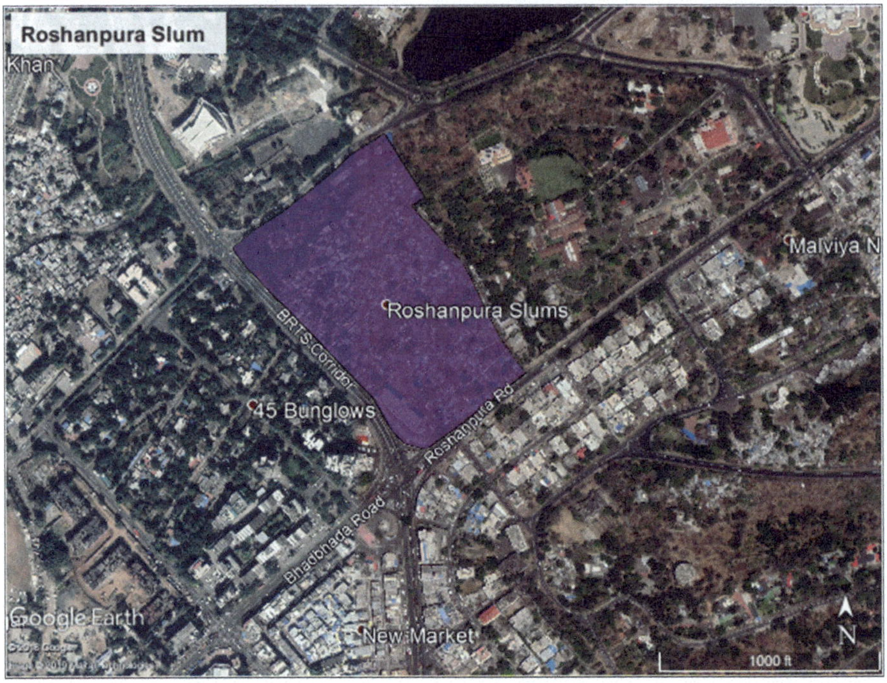


**Fig. 11** Roshanpura slum, Bhopal. *Source* Google Earth Image

collectorate. The contour levels vary from 523 to 578 m in slum, which indicate steep slopes and hence not suitable for residential purpose as majority of slum dwellings grew on undevelopable steep slope. This slum is spread over 15.7 ha area with around 450 households and accommodated more than 1750 residents. Mother India slum came in existence in 1992, and evolved on government land, reserved for majorly as green-belt (non-developable due to steep slopes) and mix-landuse along the periphery in Bhopal master plan 2005. Majority families are living in the slum from more than 25–30 years and more than 90% dwellings were in kuccha and semi-pucca condition (Fig. 14).

Based on primary survey it was found that more than 75% of population is from active working (18–60 years) age-group and were well aware about government schemes and urban poor development programs. Gondipura slum is leading in work-force participation with 82% working population, whereas it is lowest in Kohefiza slums. Income levels of all slums are ranges from Rs. 3000 to 25,000 per month, out of which 65% of HHs are earns not less than Rs. 10,000 per month. Almost 40% of households are having two wheeler ownerships and 35% are having bicycles in their possession. More than 50% population goes to work with public transport and walk in their routine practice.

Except Kohefiza slum, all other slums are more or less covered with water supply and electric services at Household level. Due to steep slopes this slum is developed on

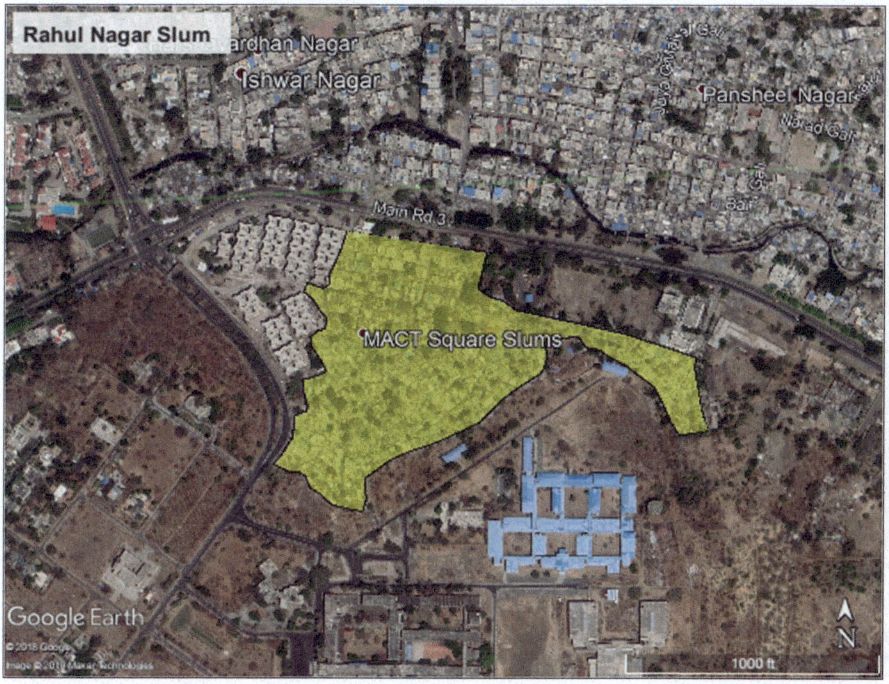


**Fig. 12** Rahul Nagar slum, Bhopal. *Source* Google Earth Image

undevelopable land, and hence not notified fully. But public stand posts are serving to all residual population not having HH level water connections in all slums. Nearly 80% of households are covered with individual toilet facility in all slums, except Gondipura, is having 60% coverage of toilets at HHs level. Rests are having access to community toilets in walkable distance, but still open defecation is practiced in all slums. Solid waste management and street lights on internal roads are major issue with all the slums. A details assessment of smart practices is done with help of use of digital instruments like; mobile phone, computers etc. to understand the role of smart technology in the life and housing conditions of slum dwellers of Bhopal city (Table 12).

Table 12 is clearly indicating that mobile phone is common article now a days and smart phone is also quite popular, especially amongst young generation. Still computer ownership and literacy is relatively low in the slums, whereas television is second most owned digital asset by slum dwellers after mobile phone. Almost all households opened their bank account under 'Jan-Dhan Yojana', but they did not link their account with mobile phone or Aadhar card. Those families, which are getting any direct government scheme benefits are, linked their Aadhar number with their respective bank account. One positive feature that emerged from the survey is initiation of digital fund transaction by slum dwellers. The young generation has

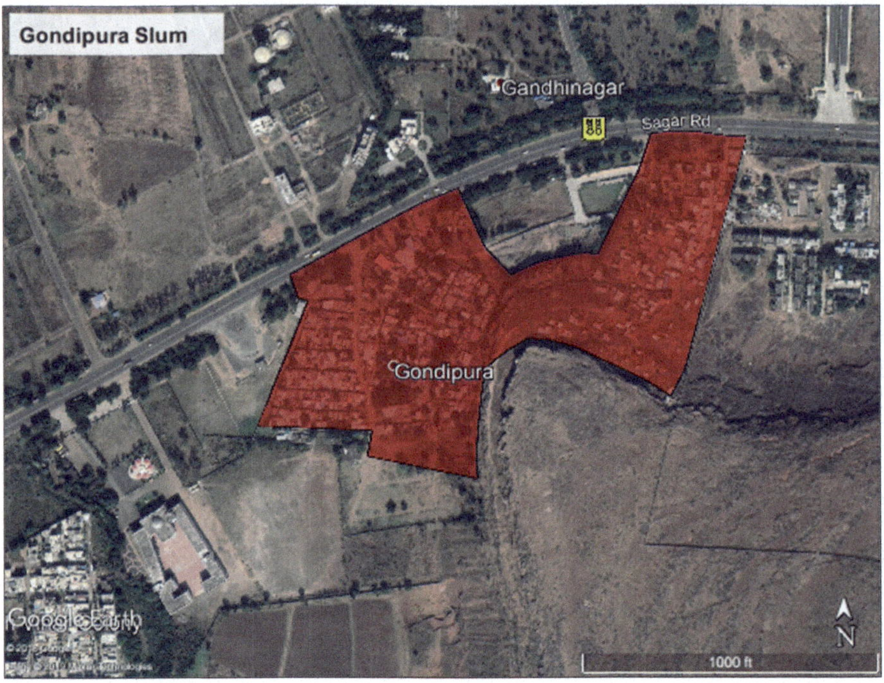


**Fig. 13** Gondipura slum, Bhopal. *Source* Google Earth Image

begun to use digital fund transfer platforms from last 3 to 4 years for their routine requirements.

Bhopal Smart City Development Corporation Limited is developing 680 state of the art government quarters near Dussehra Maidan, TT Nagar, under Area Based Development (ABD) component of the Bhopal Smart City project. The foundation stone for the Smart City government housing project was laid in December 2018 targeting inclusive housing approach for Rental, Affordable, EWS and Mixed Housing, apart from the commercial MIG and HIG housing projects. These houses will be covered with best physical infrastructure and services with long-term maintenance and management plan. With pan-city approach for smart Bhopal, other settlements clusters will be also expected to be covered with the state of art civic services by 2024, including some of the slums.

During the field investigation in the slums, it was found that most of the young slum dwellers are willing to adopt new smart services, including smart cards for public transport, smart ID (Samagra ID) for accessing government services at city level and smart fund transaction media, early information dissemination system etc. Samagra ID is a unique ID for each and every household having collective information of all the other IDs (Aadhar Card, Driving License, Passport, PAN card etc.) of all family members. People from slums are also demanding registration of their dwelling unit in the official records of BMC to avail maximum benefits from

**Fig. 14** Kohefiza slums, Bhopal. *Source* Google Earth Image

**Table 12** Ownership of digital assets and smart practices with digital technology in slums (Figures in percentage)

| HH level digital assets and practices | Roshanpura | Rahul Nagar | Gondipura | Kohe-fiza |
|---|---|---|---|---|
| Mobile phone ownership | 100 | 96 | 100 | 100 |
| Smart-phone users | 72 | 82 | 58 | 78 |
| Computer literacy | 28 | 12 | 22 | 39 |
| Computer ownership | 7 | 5 | 4 | 8 |
| Television ownership | 75 | 70 | 72 | 78 |
| Saving account in bank | 100 | 100 | 97 | 99 |
| Account linked with aadhar card and mobile phone | 41 | 26 | 30 | 34 |
| Digital/non-cash fund transfer (by PayTm, Google Tez, ATM Card etc.) | 16 | 5 | 24 | 68 |

*Source* Primary Survey

various government welfare schemes and to secure their land tenure to avoid any type of relocation from their respective dwelling units. New generation in slums is quite techno-savvy, they prefer to deposit electricity and other bills through online or mobile transactions. Use of mobile apps, sharing information with family and friends, getting awareness on various issues through smart communication sources are revolutionary practices, which are changing complete interface and life style of slum inhabitants also. Slowly these changes are percolating with aspirations of better housing conditions and quality of life in slums.

## *4.2 Informal Employment*

Bhopal has a high working rate of 29.6% out of which a total of 39% of working population is engaged in informal sector and 39% of the population is informally employed and 23% of households have their main income earners in informal sector [7]. Informal employment is seen in the form of construction workers, metal workshops, weekly vegetable/fruit markets, street vendors, repair shops and rag-pickers etc. As per a survey done by the National Association of Street Vendors in India in 2012, there were 5200 street vendors surveyed in Bhopal but actual number of vendors in the city at that time was between 15,000 and 20,000 [23]. In this study the scope of informal employment focuses on street vending activities in Bhopal city only. Within the street vending, both stationary and mobile street vendors are covered in the study.

In 2005, a scheme named 'Feri-valon-ki-Duniya' was adopted on the lines of National Policy of Street Vendors 2004. In this scheme all the surveyed and identified street vendors were given identity cards and benefits like scholarship (for 2 children), health assistance, cash for daughter's wedding, Rs. 30,000 to the family on death of a vendor along with license transfer to a family member and pregnancy and delivery assistance to wives of street vendors (6 month wages to wives and Rs. 1000 for delivery in government hospital) [23].

Street Vendors (Protection of Livelihood and Regulation of Street Vending) Act, 2014, has been enacted to protect the rights of street vendors including social protection, but most of its norms and guidelines are not enforced at ground level in Bhopal City. Constitution of street vending committee and demarcation of ward wise vending zones are still pending, whereas registration of vendors was started and is still in progress with Bhopal Municipal Corporation (BMC). It is proposed to provide Vendor ID Cards to street vendors by BMC to effectively execute betterment efforts.

To understand the impact of some of the smart practices on urban informal employment, four prominent markets are selected for the study purpose. These are as: New Market, Karond market, Taj market and Lalghati square. Total 200 samples were collected from all four selected markets through stratified proportionate random sampling technique, out of which 70 were mobile street vendors and 130 were stationary street vendors during primary surveys (Fig. 15).

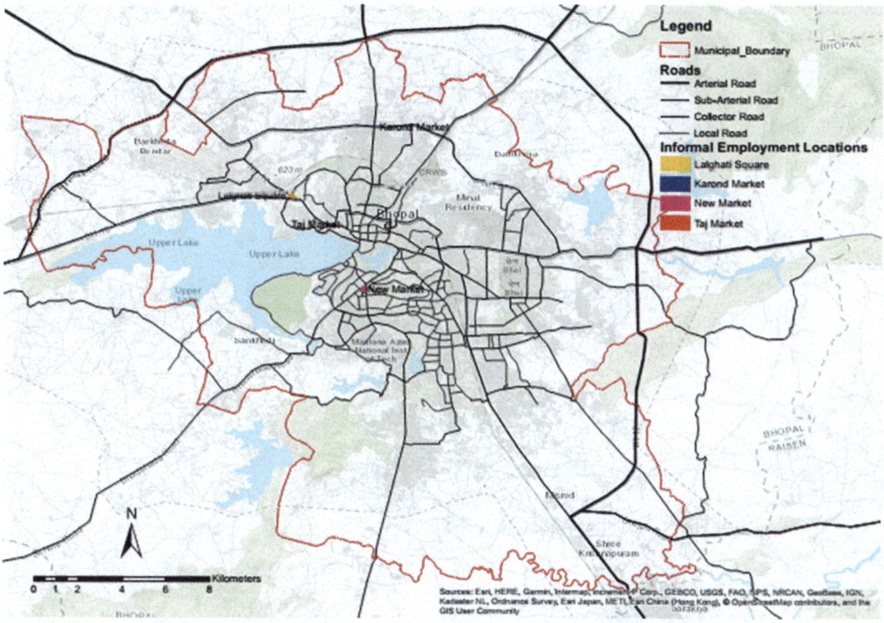


**Fig. 15** Location map of selected market areas for the study of informal employment in Bhopal *Source* School of Planning and Architecture [35]

'New Market', one of the main business district of Bhopal, located in the center of the city, includes shops, offices and various formal and informal trade establishments. Informal employment in this area is mainly from the street food and fast food stalls, and other types of goods sold mainly comprises of garments, bags, kids toys, balloon sellers and artificial jewelry products etc. Both mobile and stationary vendors are operating from new market, but the share of representation for informal employments, based on primary survey, is nearly about 35%. 'Tin shed' area located to the south of New Market is a designated space for informal vendors buzzing with variety of commercial activities. Items like handicrafts, woolen and other clothing items, bamboo articles, bangles and general items can be bought from the shops and vendors here. The nature of majority of shops here is stationary (60%) and rest is mobile (Fig. 16).

'Karond Market' is one of the prominent commercial places in north Bhopal, which is connected to the airport and Bhopal IT Park, Berasia road, Ayodhya Nagar Bypass towards Raisen city. This market has almost equal representation of informal and formal commercial activities. Informal employments are practiced on and along the streets of Karond square and these vendors are trading in vegetables, fruits, eateries, kitchen items etc. mostly at small scale (Fig. 17).

**Fig. 16** New market, Bhopal. *Source* Google Earth Image

'Taj Market' is located on Sultania Road, near Taj-ul-Masjid (biggest Masjid in Bhopal and Central India), opposite to Gandhi Medical College and near to Hamidia Hospital complex. This strategic location offers a variety of commercial activities in formal and informal sector here, out of which shops of informal nature includes; fruits, flowers, eateries stall etc. On every Friday, a dedicated informal market (weekly market) offers a variety of goods to people here, targeting especially low income groups and weaker sections (Fig. 18).

Lalghati square area is the prime junction of Bhopal city connecting Airport road, Bairagarh sub-city, VIP Road (connecting central city area), Sultania Road (connecting District Collectorate and other government offices) and Gufa Mandir Road. Therefore, it is also a hub for the sector of informal employment in Bhopal. The average age of the vendors engaged in this area for informal employment is around 38 years where majority of the share is of males. Trading is done in eatables such as vegetables, fruits, paan, etc., mostly at small scale, where the paan shop contributes a major share. This is evident from the fact that most of the sellers who come here are educated only up to primary level or/and most of them are illiterates (Fig. 19).

Around 20–30% of the vendors are illiterates and 40–60% of vendors have not studied beyond high school from the selected samples in different markets. Vendors of New Market have better education level as compared to other markets, whereas

**Fig. 17** Karond market, Bhopal. *Source* Google Earth Image

people at Karond Market are from the poorest education profile. Around 60% of vendors change their commodity types with changing weathers and demands. Vendors from Lalghati Market are more stable in terms of the products sold, whereas vendors from Taj Market frequently change their goods. Majority of vendors in the market areas are stationary in nature, but Karond Market is dominated by mobile street vendors.

More than 80% of the vendors are aware about the cleanliness and, the vendors and the shopkeepers of each particular market areas collectively contribute to maintain cleanliness by paying the municipal sweepers. Almost 60% of the street vendors arrange water and electricity from nearby shops or residences. They are also aware of making complaints for solid waste management and civic services to BMC but at the same time they are afraid of relocation or removal from specific location of their routine business. Therefore in any case, if action is needed, they refrain from approaching the local officers and prefer to manage at their own level through known sources. Security of street vendors is also one of the crucial aspects as they are one of the most vulnerable groups and hence BMC does its duty by maintaining the security of the area by installing and maintaining street lights and police patrolling is also ensured by city police, which has led to reduction in the instances of harassment and looting.

**Fig. 18** Taj market, Bhopal. *Source* Google Earth Image

Nearly half of the vendor population prefers two-wheelers for their daily commuting while others are using public transport and/or walk for their daily mobility. Majority of the vehicles of the vendors are parked on the streets. They park their vehicles anywhere near their shops or stalls. Almost no vendor in the New Market area is willing to use the smart parking facilities due to the higher parking charges, which they feel unaffordable. A detailed assessment is undertaken for their smart practices with the help of digital instruments like: mobile phone, computer, etc. to understand the role of smart technologies in the life of street vendors.

Table 13 clearly indicates that due to the availability of gadgets at affordable prices and convenient existence of second hand smart phone market, more or less all the sellers and the street vendors own mobile phones. But understanding and use of computers is seen negligible as compared to very good access to mobile phones. On an average around more than 75% of the vendors own smart mobile phones and nearly 60% of them are using various applications like WhatsAapp and Facebook, etc., whereas remaining 40% use the standard 2G mobile sets. This situation of mobile literacy is more or less same for vendors below 35 years of age in all study locations.

**Fig. 19** Lalghati market, Bhopal. *Source* Google Earth Image

**Table 13** Ownership of digital assets and smart practices with digital technology by street vendors (Figures in percentage)

| HH level digital assets and practices | New market | Karond market | Taj market | Lalghati square |
|---|---|---|---|---|
| Mobile phone ownership | 100 | 100 | 93 | 100 |
| Smart-phone users | 85 | 75 | 68 | 80 |
| Smart apps regular users | 65 | 56 | 57 | 70 |
| Computer literacy | 8 | 20 | 10 | 12 |
| Use of apps/internet for procurement, marketing, complaints, awareness | 32 | 27 | 20 | 24 |
| Saving account in bank | 83 | 74 | 66 | 78 |
| Account linked with aadhar card and mobile phone | 35 | 50 | 23 | 68 |
| Digital/non-cash fund transfer (by PayTm, Google Tez, ATM Card etc.) | 30 | 27 | 17 | 23 |
| Use of smart services (parking, municipal charges) | 8 | 9 | 6 | 5 |

*Source:* Primary Survey

## 5 Smart Governance

The Bhopal Municipal Corporation (BMC) has been playing the role of custodian of civic affairs in Bhopal city under the provisions of the Madhya Pradesh Municipal Corporations Act. The total area under BMC is 648.24 $km^2$. The city is divided into 85 wards. Each ward elects a councillor. The Mayor of Bhopal is currently elected directly. The Municipal Commissioner is the highest officer of BMC, which is responsible for the departments of public works, water supply, planning and development, fire brigade, health and sanitation, etc.

Giffinger's model [12] for analyzing Smart Governance in the context of Smart City included: Participation in decision-making; Public and social services; Transparent governance; Political strategies and Perspectives. Cohen's smart city model [11] gives characteristics similar to R. Giffinger's, but the factors are different, with more strongly expressed digital dimension. This model has separate factors, such as Open data, ICT and e-government, Integrated ICT [12].

Praharaj et al. [32] reviewed the performance of Indian smart cities in attracting citizens to engage in discussions on 'MyGov.in' platform. Among six different activities hosted in 'MyGov.in' website, the research focused on two components: (1) number of comments that were observed in city specific discussion on smart city proposals and (2) number of essays submitted by citizens on the platform to share their vision of smart city. The intensity of citizen engagement was found to vary greatly across 89 cities that invited discussions on smart city proposals. Whereas seven cities recorded participation of more than 0.1 million citizen comments, 28 cities did not even attract 100 participants on the same platform. It was found that Bhopal attracted the maximum number of discussion (0.17 million). 70 out of the 100 nominated smart cities in India accepted essay submissions from its citizens describing smart city vision. Bhopal came out on top in number of essay submissions with 5091 entries, doing well in drawing citizens into discussions also.

The parameters considered for detailed investigation by Praharaj et al. [32], for which data was collected from 89 cities, included: Population participated on MyGov.in (%); Households having computer/laptop with internet (%); Households having mobile phone (%); Literacy Rate (%); Work participation rate (%); and, Service sector workers (%) The interesting observations include: a negative correlation between access to internet and intensity of eParticipation; negative correlation between ownership of mobile phones and eParticipation on MyGov.in; and, negative correlation between literacy rate and eParticipation. The study concluded that the awareness and interest of citizens in engaging with governments and range of stakeholders plays a larger role than their abilities and digital skills, and that smart cities of the future should focus on digital initiatives based on objective local assessments of citizen's behaviours and aspirations.

Information and communication technologies (ICT) maturity is playing a significant role in the achievement of global competitiveness at the country and also at the city level. The following sections investigate the socio-economic and technological factors that are most likely to be associated with ICT maturity. To understand the socio-economic profile of the city three separate datasets for wards in BMC from published data of Census 2011 are used. The first dataset used is that of percentage of households having temporary roof for all wards of BMC. As the latest population data is from Census 2011 and number of wards in BMC has increased after 2011, there are a few wards for which the information which information are not available. It is observed that, wards like 50, 54, 60, 57, 66 and 68 have very high percentage of households living in houses with temporary roof. Based on this observation, these wards lying in the south west of BMC can be assumed to have relatively low income profile. From primary survey it could be understood that these wards are inhabited by large numbers of daily wage earners who belong to the low income group. The second dataset used is that of percentage of households having no house or house with only single room. This information are available all wards of BMC except the wards created after 2011 (Fig. 20). It is observed that, wards like 50, 54, 60, 57, 66 and 68 have very high percentage of households with no house or houses with only one room. Based on this observation it can be assumed that these wards are inhabited by low-income groups. There is a strong similarity between households having temporary roof and households having no house or house with only single room. These two parameters will help to further understand the spatial variation in socio-economic condition across BMC. The third dataset used is that of percentage of households having semi-permanent or temporary structure. It is available for all wards of BMC except the wards created after 2011. It is observed that, wards like

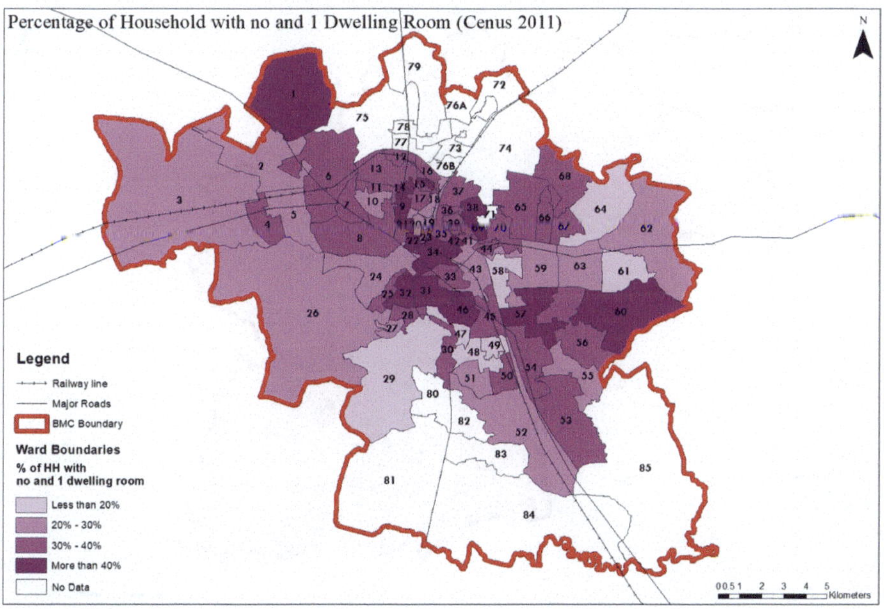


**Fig. 20** Percentage of households with no or 1 dwelling room in Bhopal city. *Source* Census of India [9]

56, 60, 54, 50, 57, 52, 53 and 68 have very high percentage of households with semi-permanent or temporary structure. Based on this observation it can be assumed that these wards are inhabited by low-income groups. Thus, using triangulation, it can be said with a relatively high degree of confidence that this particular group of wards have a high concentration of the urban poor and socio-economically disadvantaged. Now regarding the ICT maturity and to measure the efficiency of governance-related apps launched in the Smart City era in BMC, data has been collected regarding the usage of the two apps which are Bhopal Plus and Mayor Express. The data collected from BSCDCL office in March 2019 includes the number of people who are using a particular service in the app in each ward of BMC.

According to the latest published information on the www.smartcitybhopal.city portal: "Bhopal Plus serves as a single digital platform for the citizens to engage with the government, avail services, receive issue resolution, and learn about Government schemes, projects, and initiatives [28]. The four main components of the platform are Citizen Collaboration, Grievance Redressal, Citizen Service Delivery G2C and B2C services) and Public Dashboard. Revolutionizing people's connect with the Government and services, the App provide instant, electronic access to the information people need, the services they require, and the interaction they want." For the present study, the usage levels of Bhopal Plus app were mapped for each ward of BMC (Fig. 21). It was observed that penetration of the app is very low up to 6% in wards like 54, 56, 60 and 52. These wards were also observed to have high

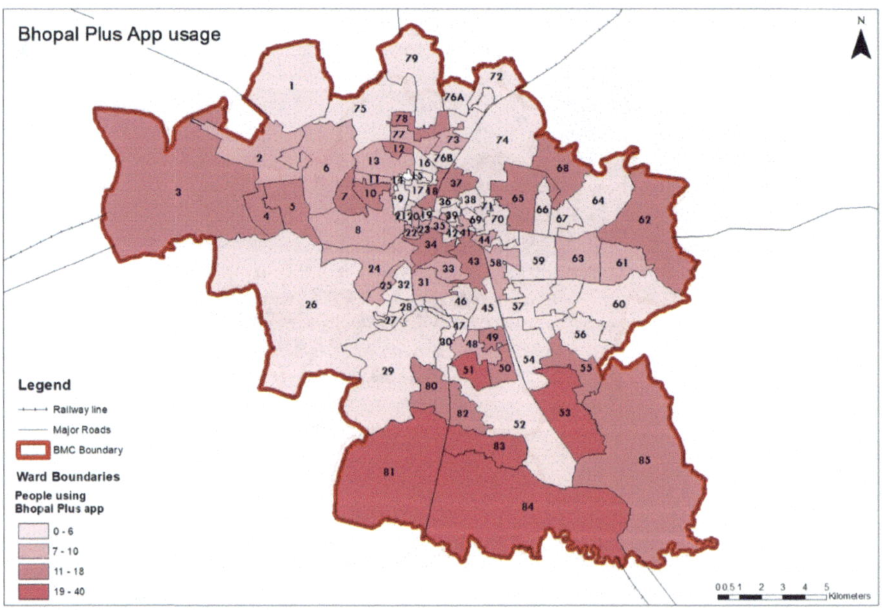


**Fig. 21** People using Bhopal Plus app in Bhopal city. *Source* BSCDCL [6], Singh [38]

proportions of low-income households in the previous analysis. Thus a case can be made that the digital urban governance innovations lack inclusivity in Bhopal.

According to latest published information on the portal https://mayorexpress.in/: "Mayor Express provides reliable and affordable home based Services and Service Providers" like Plumber, Electrician, Carpenter, Gardener, Driver, Mason, Painter, Sweeper, Labour, Barber at Home, Beauty Salon at Home, Sofa Repair, Car Wash, Appliances Repair, RO Services, Pest Controls, AC Repair and Servicing and other home services [29]. All Services can be availed easily through Call Center and Bhopal Plus App, available on Android and iPhone Platforms. Usage levels of Mayor Express app were also mapped for each ward of BMC (Fig. 22). The overall penetration of Mayor's App was seen to be low. The maximum usage level of this in a ward was just 13% till the time of data collection, which was found to be very less as under Smart City initiative, a target of 50–70% usage of the app was learnt to have been set. It was learnt from primary surveys and interviews that although the grievances registered in the app are solved effectively but there is low usage of this app. However, it may not be directly attributable to socio-economic factors. Rather the services offered through the Mayor's App are also easily available at community level in Bhopal and hence the preference for this app may be relatively low.

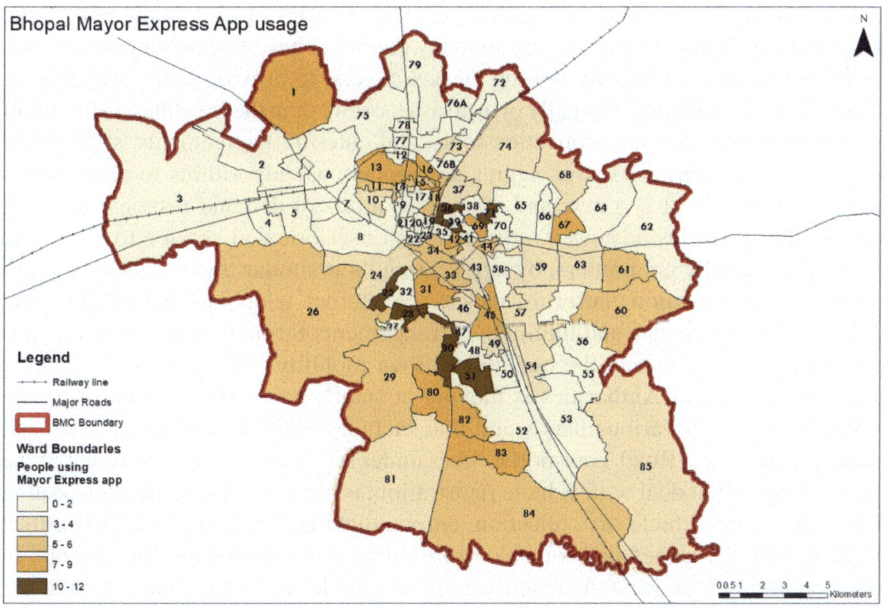


**Fig. 22** People using Mayor Express app in Bhopal city. *Source* BSCDCL [6], Singh [38]

## 6 Conclusion

Gradual change is visible in the field of Public Transport Development accessibility improvement measures in Bhopal with NUTP 2006 and JnNURM, and Smart Cities Project. However, to keep the pace of development, there is a need to adopt many innovative strategies. Most of the developments are still concentrated towards solving the problems from the point of view of private vehicle users. Many consider public transport bus service as an inferior 'public good'. Based on the income change, the perceived value of time safety, quality, and expectations of the comfort would change. Public transit agencies are delivering transport services in a rapidly changing and highly competitive transportation market. For ensuring the rider's satisfaction in such an environment, it is essential to conduct market segmentation studies to adopt different marketing strategies. Market segmentation analyses are commonly employed by transit agencies to identify groups of users, which are subsequently used as a base for developing policies and strategies aimed at improving customer satisfaction. Buses with different hierarchies of value-added services and in-vehicle services for the same route should be added to the system. Frequency and adequate space availability in public transport is an important factor in attracting public transport patronage. In Bhopal, almost all the routes have headway of 10 min.

The existing public transport models are rigid in terms of route, time table, fare, stoppages, etc. Technology start-ups in many countries have begun to aggregate privately owned buses, typically seating 20 passengers, to provide transport services. These "Bus Aggregators" typically own no assets other than the Information Technology infrastructure and fare-collection tools, and enter into agreements with private bus owners for operations. Bus aggregator models use algorithms to chart routes based on demands. It is a new and innovative mode of demand-responsive transit systems that provides advanced, attractive and user-oriented service to a specific clientele, especially commuters, by aggregating their similar travel-demand pattern using online information platforms, such as the internet, telephone and smart phone.

Transport governance and intuitional arrangements for managing various modes of transport will influence the quality of urban mobility. The creation of Unified Metropolitan Transit Authorities is moving at snail's pace. Management of transportation data is in various hands and that includes central, state (province), and municipal agencies. Road Transport Offices under the state government is one of the nodal agencies that deal with vehicle registration, issuing of driver's license, issuing of permit, motor vehicle tax collection, enforcement etc. A Road Transport Office Division boundary does not coincide with Urban Area Boundary. This results in issues regarding/like accurate quantification of vehicles available in an urban area.

Cities in the global north have agencies that are rapidly adopting General Transit Feed Specification (GTFS) as the de facto standard of storing and disseminating transit schedules and associated geographic information. It helps fusing of transit information to platforms like Google Maps. There is an astonishing boom in the fast-growing transit data sector. More and more authorities and companies are opening up their data in order to share the benefits of transit analysis. However, in the case of cities like Bhopal in India, the availability of data is itself a considerable challenge. Very few public transit agencies work with spatial data, and even fewer publish such data. Most agencies do not have the in-house expertise to handle transit data. Open data ecosystems are more likely to encourage innovation and give a fillip to new ideas, tools, and approaches.

Intermediate Transit, Para Transit, Taxi services as well as auto- and cycle- rickshaws that ply on shorter routes are also common. Multimodal integration allows a transfer between different modes of transport to become easy and seamless. It plays an essential role in inducing the use of public transport, as a commuter's travel choices are made based on their relative convenience for the entire trip, and not just a segment of that trip. However, the physical and fare integration between these modes are not properly implemented. Each of these modes works independently and competes for each other. With the introduction of BRTS services in Bhopal, many IPT and Para-transit routes of city services are overlapping. It leads to incoherent schedules, overlapping routes, time tables, and a higher interchange penalty. IPT and various Para Transit mode routes should be re-assigned to provide last-mile connectivity to improve the access and egress trips. Fare integration can create better patronage to public transport and save the time for commuters. Lack of fare integration is also forcing commuters to pay more for the trip.

Bus Rapid Transit is low-cost, bus-based public transport mode that echoes the standards of some of the rail based mobility systems. BRTS has segregated right of way, pre board ticketing, and closed station. It places the low-cost high occupancy transport in a genuine democratic road space. BRT is a low-cost and equally efficient public transport system in the world. Though 10 corridors were proposed initially, only one corridor was implemented in Bhopal. The rest of the corridors may be implemented for better accessibility.

Regardless of how fuel-efficient the cars are or how little pollution they emit, cars need to be parked somewhere, and the average car spends about 95% of its life parked [37] and uses several parking spaces each week [18]. Parking standards need to be amended by incorporating parking supply caps, parking maximums and flexible parking standards. Public transport accessibility, availability of mixed land use, bicycle-sharing facilities, walkable streets, residential density, employment density, etc. should be considered in deciding the parking standards. Considering the bigger role of parking charge in the average generalised cost of trip, private parking supply should be adequately charged, and the concept of free parking at the workplaces should be abolished. Legal tools should ensure the availability of parking spaces or ownership certificates of parking spaces for vehicle registration.

The new generation is willing to explore internet/apps, for procurement of goods at a cheaper cost without compromising quality, for the marketing of their products, for getting the latest information related to their trades and for posting their issues and complaints to the concerned authorities. Almost one-fourth of vendors with smartphones have started using internet applications for the above said practices. But the use of smart parking and smart payment process for municipal charges is still not popular. The majority of vendors have their individual savings bank account, but a lot of them have still not linked their account with Aadhar card and mobile phone, and that's why the share of on-line fund transactions is quite less. Almost none are aware of the micro-finance institutions. Only 10% of the shopkeepers and vendors are regular tax-payers. After linking it with the socioeconomic profile of wards and the overall usage of digital governance apps, it is seen that the apps are not probably reaching out to the lower-income groups effectively, pointing towards a lack of inclusivity. Ownership of digital assets is quite high among slums and relatively low-income people, but still, the usage of Mayor Express and Bhopal plus apps are quite low in those wards dominated by low-income people. Bhopal Livability indicators are getting improved year by year with various changes in public transportation, bicycle- sharing facilities, management of parking facilities, and overall traffic management. Bhopal city has taken up many projects for sustainable mobility promotion and improvement of the city livability.

**Acknowledgements** We acknowledge the support received from BMC, BSCDCL officials regarding sharing of data, information and documents. We also thank SPA Bhopal B. Plan 2016–2020 batch students, Mr. Sumit Singh (MURP, 2018 batch of SPA, Bhopal) and Ms. Sakshi Khare (Masters student of SPA, Bhopal), who put their best efforts to make this research possible.

## References

1. Arora S, Khare R, Saxena S (2016) Morphological patterns of dwelling units in informal settlements of Bhopal, India. J Civ Eng Environ Technol 3(9):815–819
2. BCLL-Bhopal City Link Limited (2018) BCLL built for Bhopal—an urban transit company. Bhopal City Link Limited. http://mybusbhopal.in/. Accessed 24 Nov 2019
3. BMC-Bhopal Municipal Corporation (2014) Bhopal parking master plan. Revised final report. Bhopal Municipal Corporation, Bhopal
4. BMC (2015) Bhopal parking master plan. Revised final report. Bhopal Municipal Corporation, Bhopal
5. BMC (2012) Public bike sharing schemes for Bhopal city, Final Detailed Project Report, Bhopal Municipal Corporation, https://smartnet.niua.org/sites/default/files/webform/PBS%20Bhopal%20DPR_23%20JUNE%20(2%20files%20merged).pdf. Accessed 24 Nov 2019
6. BSCDCL-Bhopal Smart City Development Corporation Limited (2019) Request for proposal for selection of implementation agency for Integrated Traffic Management System (ITMS) in Bhopal City. Bhopal Smart City Development Corporation Limited. https://smartnet.niua.org/content/aa5755c1-74d5-4063-b1c4-6f152a9c4e2e. Accessed 24 Nov 2019
7. CDP (2006) Bhopal city development plan. Bhopal Municipal Corporation. http://www.mpurban.gov.in/Pdf/CDP/Bhopal%20CDP_Final%20.pdf. Accessed 24 Nov 2019
8. CMP (2014) Comprehensive mobility plan. Bhopal Municipal Corporation, Bhopal
9. Census of India (1951–2011) Primary census abstract. Office of the Registrar General & Census Commissioner, Government of India
10. Chen M, Bonner C, Carre F (2015) Organizing informal workers: benefits, challenges & successes. Background paper. UNDP Human Development Report Office, New York. http://hdr.undp.org/sites/default/files/chen_hdr_2015_final.pdf. Accessed 24 Nov 2018
11. Cohen B (2013) Boyd Cohen: "the smart city wheel". https://www.smart-circle.org, https://www.smart-circle.org/smartcity/blog/boyd-cohen-the-smart-city-wheel/. Accessed 10 Nov 2019
12. Giffinger R, Fertner C, Kramar H, Kalasek R, Milanovic NP, Meijers E (2007) Smart cities—ranking of European medium-sized cities. Vienna University of Technology, Vienna

13. International Labour Organization (2018) World employment social outlook: 2018 trends. International Labour Organization, Geneva. https://www.ilo.org/wcmsp5/groups/public/—dgreports/—dcomm/—publ/documents/publication/wcms_615594.pdf. Accessed 24 Nov 2018
14. KPMG (2018) Green and blue master plan for Bhopal smart city—draft green master plan. Bhopal Smart City Development Corporation Limited, Bhopal
15. Khan MS (2004) The Begums of Bhopal—a dynasty of women rulers in Raj India. Viva Books Private Limited, New Delhi
16. Kumar Y, Singh S (2014) Rehabilitation of slum dwellers: a case study in Bhopal. Soc Change 44(4):567–578. https://doi.org/10.1177/0049085714548540
17. Lennard SC, Lennard H (2008) The principle of true urbanism. http://www.livablecities.org/articles/principles-true-urbanism. Accessed 19 Apr 2012
18. Litman T (2013) Parking management: strategies, evaluation and planning. Victoria Transport Planning Institute. http://www.vtpi.org/park_man.pdf. Accessed 15 Mar 2014
19. MOSAPI-Ministry of Statistics and Program Implementation (2018) India in figures 2018. Report of Social Statistics Division, Central Statistics Office, Ministry of Statistics and Program Implementation, Government of India
20. MPMRCL—Madhya Pradesh Metro Rail Company Limited (2019) Environmental impacts assessment and environmental management plan of Bhopal metro rail project. Madhya Pradesh Metro Rail Company Limited, Bhopal, Madhya Pradesh. http://www.mpmetrorail.com/files/Final%20EIA%20&%20EMP_October%202019.pdf. Accessed 24 Nov 2019
21. Mahadevia D, Joshi R (2009) Subversive urban development in India: implication on planning education. Center for Urban Equity, CEPT University, Ahmedabad
22. Mishra S (2019) Bhopal: with ITMS, e-challan goes hi-tech but fine collection is still low key. Times of India, 21 Aug 2019. https://timesofindia.indiatimes.com/city/bhopal/with-itms-e-challan-goes-hi-tech-but-fine-collection-is-still-low-key/articleshow/70761347.cms. Accessed 24 Nov 2019
23. Mitra T (2017) Street vendors the powerhouse of Bhopal, from urban sketchers. http://www.urbansketchers.org/2017/10/street-vendors-powerhouse-of-bhopal.html. Accessed 24 Nov 2018
24. MoHUPA—Ministry of Housing and Urban Poverty Alleviation (2009) India urban poverty report 2009. Ministry of Housing and Urban Poverty Alleviation, New Delhi
25. MoRTH—Ministry of Road Transport and Highways (2017) Road accidents in India—2017. Ministry of Road Transport and Highways. http://www.indiaenvironmentportal.org.in/content/459084/road-accidents-in-india-2017/. Accessed 24 Nov 2019
26. MoUD—Ministry of Urban Development (2016) Liveability standards in cities. Ministry of Urban Development. http://smartcities.gov.in/upload/uploadfiles/files/LiveabilityStandards.pdf. Accessed 24 Nov 2019
27. NCAER—National Council for Applied Economic Research (2013) Annual report of national council for applied economic research (NCAER), New Delhi
28. Official website of 'Bhopal Smart City Development Corporation Limited'. www.smartcitybhopal.city. Accessed 2 Nov 2019
29. Official website of 'Mayor Express' app. https://mayorexpress.in/. Accessed 2 Nov 2019
30. Pandey RU, Garg YK (2015) Evaluating livability performance of residential built environment: a step towards making a smart city. Spandrel 11:39–51
31. Planning Commission (2012) Working group on urban poverty, slums, and service delivery system, New Delhi
32. Praharaj S, Han JH, Scott H (2017) Innovative civic engagement and digital urban infrastructure: lessons from 100 smart cities mission in India. In: International high-performance built environment conference—a sustainable built environment conference 2016 series (SBE16), Elsevier, Sydney, pp 1423–1432
33. Roy A (2005) Urban informality: towards an epistemology of planning. J Am Plann Assoc 71(2):147–158
34. Saxena PN, Joshi R (2015) Eradications of slums in Bhopal city. Int J Sci Technol Eng 2(6):105–112

35. School of Planning and Architecture Bhopal (2018) Local area plan for smart city area, Bhopal (Unpublished)
36. Sengupta AK (2009) The challenge of employment in India: an informal economy perspective. Report of the National Commission for Enterprises in the Unorganised Sector, Government of India, Volumes 1 and 2, Academic Foundation, New Delhi
37. Shoup DC (2005) The high cost of free parking. American Planning Association, Planners Press American Planning Association, Chicago
38. Singh S (2019) Assessment of socio-economic inclusivity in smart city Bhopal. Dissertation. School of Planning and Architecture Bhopal (Unpublished)
39. ToI—Times of India (2019) Bhopal: no end to parking woes in sight. Times of India, 21 Aug 2019. https://timesofindia.indiatimes.com/city/bhopal/no-end-to-parking-woes-in-sight/articleshow/70761601.cms. Accessed 24 Nov 2019
40. United Nations (2014) World urbanization prospects: the 2014 revision. Department of Economic and Social Affairs, Population Division, CD-ROM Edition
41. VTPI—Victoria Transport Policy Institute (2010) Transport strategies and plan department of transport. State Government of Victoria, Australia. http://www.transport.vic.gov.au. Accessed 24 Nov 2018

# Geospatial Intelligence for Smart Living–Case of New Delhi

**Kusum Lata, Priyanka Kumar, and Arpita Banerjee**

**Abstract** The following chapter tries to pay attention to the spatial dimension of smart living projects in India. Geospatial Planning is a comparatively new terminology in the spatial planning dictionary of the Indian urban scenario, just like the terms smart planning and smart cities are. Although there is no clear delineation between smart living and smart city projects in India, characteristics of the Smart Cities Mission are in consonance with the ideologies of smart living. The chapter first tries to understand the key values underpinning the concept of smart living and delineates it through the terms interconnectedness, sustainability and mobility (both social and physical in nature). It tries to then identify how geospatial tools help manifest these values of smart living and an elaborate network of indicators is identified. This network of indicators lays the foundation for evaluating geospatial smart living projects in New Delhi.

**Keywords** Smart geospatial living · Smart city · Urban planning · New Delhi Municipal Corporation (NDMC) · Delhi Spatial Data Infrastructure (DSSSI)

K. Lata (✉)
Centre for Urban Studies, Indian Institute of Public Administration, Room No. 35A, First Floor, I.P. Estate, Ring Road, New Delhi 110002, India
e-mail: kusumlata.iipa@iipa.org.in; drkl.iipa@gmail.com

P. Kumar
Regional Centre for Urban and Environmental Studies (RCUES), Lucknow, India
e-mail: priyanka.kumar275@gmail.com

*Present Address:*
#1139 New Housing Board, Karnal, Haryana 132001, India

A. Banerjee
Department of Physical Planning, School of Planning and Architecture, New Delhi, India
e-mail: arpitab.iipa@gmail.com

*Present Address:*
c/o Shri Arup Kumar Banerjee, Saptashwa, Near Parijat Apartment, Hindustan Park, Paschim Bardhaman, Asansol, West Bengal 713304, India



T. M. Vinod Kumar (ed.), *Smart Living for Smart Cities*,
Advances in 21st Century Human Settlements,
https://doi.org/10.1007/978-981-15-4603-7_4

Before the analysis of geospatial intelligence for smart living in New Delhi, a brief description of the evaluation area, its demographics, spatial and administrative attributes are laid down. Afterwards the geospatial infrastructure underlying smart living projects in New Delhi are analysed on the basis of the evaluation criteria that was assembled from the literature. The chapter ends with the recognition of major gaps in the currents status of geospatial smart living and its conception in New Delhi Geographic Information System),GLONASS

**Aim, Objectives and Methodology**

**The aim of the study is to understand Smart Living in New Delhi through the use of geospatial tools and technologies adopted to enhance the quality of life of its citizens and their impacts on the development of New Delhi**.

Objectives of the study are

- To understand what is a Smart Living and the role of Geo-spatial Tools and Technologies in reinforcing it.
- To analyze the different components which constitute smart living in NDMC Area, and asses and evaluate their geo-spatial nature.
- To identify present issues of Geospatial Interventions in Smart Living and propose broad improvements.

The literature study is divided into two broad parts. The first deals with the nature and concept of allied terminologies such as smartness, smart living and smart cities. Components of geospatial smart living will be identified and a detailed evaluation criteria for geospatial intelligence in smart living will be handled in the subsequent section. New Delhi Municipal Council (NDMC) Area is introduced and analyzed. After a cursory glance at different Smart Living projects in the NDMC Area important geospatial interventions shall be identified for in-depth study. These projects will be utilized for in-depth analysis and evaluated based on the components of smart living identified earlier. Post analysis, the lacunae in the existing scenario of geospatial smart Living in NDMC area will be identified. (the scope of geo-spatial tools will be carefully delineated and its subsequent utilization in NDMC area will be critiqued). Policy recommendations at the general level will be provided (Fig. 1).

## 1 Introduction

Smart cities represent the zenith of modern-day urbanization and typically emerge from the forces of urbanization. Smart cities can also be visualized as a solution to the problems of urbanization and population growth, creating dense and compact civilizations which are livable, sustainable as well as resource-efficient. Multiple roles of a smart city require collaborative inputs from multiple disciplines spanning planning, architecture, transportation, engineering, geography, social sciences, management, finance and policymaking (Harrison and Donnelly [13]). A smart city is

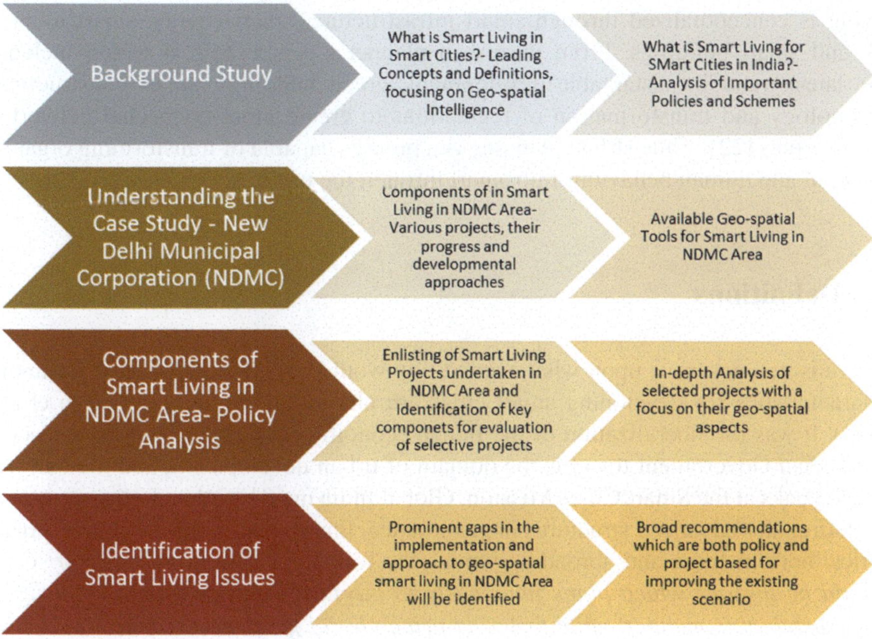


**Fig. 1** Methodology of the study

usually a global, interconnected city within a strong network of well-coordinated communication systems driven by ICT which creates horizontal links between cities of the same hierarchy within the global urban ecosystem and enables a platform for a knowledge-based world economy. Hence, smart living is an interconnected, webbed digital medium of living that enables a constant flow of ideas, experiences and Spatio-temporal characteristics, making the global, local. Within a smart city, technology reduces the friction of distance by cutting costs and making long-distance travel faster and more accessible [29]. Smart living in smart cities would, therefore, be characterized using terms like intelligent, wired, technological, digital, knowledge-based ad creative. Smart cities also envision improvements in politics and democracy through the use of ICT, making it their primary characteristic. Other characteristics of smart cities that have an impact on smart living include creative entrepreneurship and business-driven urban development, inclusion, and equity, the building of social capital and sustainability.

Thus smart living is a much more pervasive concept than smart cities. Although in India, the basic mechanism for driving smart living is through the concept of a smart city. In smart living, as the European Commission puts it, it involves transformations in the standard of living and quality of life of a city's inhabitants [26]. It affects how and where people live, work and travel within and outside the city while maximizing their socio-economic prosperity and environmental sustainability. Smart living is thus a form of a better, more balanced lifestyle. In smart cities, smart living is conceptualized through smart infrastructure, smart energy, smart mobility and smart buildings. From the urban planner's perspective, it would include the latest trend in sustainable development, green buildings, smart construction technology and transformation of regulations to create more connected networks across cities [22]. Thus Urban planning is a process capable of transforming organizational and human behaviour into smart through the use of smart infrastructure.

## 2 Definitions

There is no consensus upon what a smart city would be, although several common characteristics for developing smart cities are ubiquitous throughout (Khan et al. [15]). It was the liberalization of the Indian economy, that centralized the focus of the Indian Government towards the domain of urban development, and the current graph peaks at the Smart Cities Mission. Global India must be urban India, equipped with digitally advanced structures in governance, finance and public welfare. Smart cities, both in India and abroad epitomize the following definition "*A Smart City would be the one which plans judiciously to meet its aspirations and challenges in a sustainable manner while fostering principles of good governance. These are achieved in a Smart City by utilising the enhanced power of technology, engaging with a more aware and informed citizenry and creating a more competent and capacitated set of people working within an accountable framework*" (Banerjee and Rathi 2015).

The smart city must inculcate competencies on the part of the government, the people, and their planners to integrate the two banks of a hiatus, technology and urban planning, and do so in a manner that is effective, efficient and sustainable in nature.

### *2.1 Smart City Mission in India*

The *Smart Cities Mission (SCM)*, a flagship scheme under the Ministry of Housing and Urban Affairs of the Government of India (GoI), was launched in India in June 2015, with a whopping fund to the tune of Rupees 48,000 crore (14 billion USD) to build a 100 smart cities which were selected in four phases from across the nation (Fig. 2) (GoI 2015).

Smart cities are conceived as a technological revolution that would amplify the development of urban India by notches that would compete globally. The Smart Cities Mission in India combines the advantages of digital planning with urban development in an attempt to solve what is one of the biggest crises of urbanizing third world countries—unregulated and unsustainable city growth (Datta 2015).

As Bhattacharya and Rathi had shown in 2015, there is a need to reconceptualize the Indian understanding of smart cities by learning from the Millennium and Sustainable Development Goals (MDGs which have later translated into SDGs). Smart

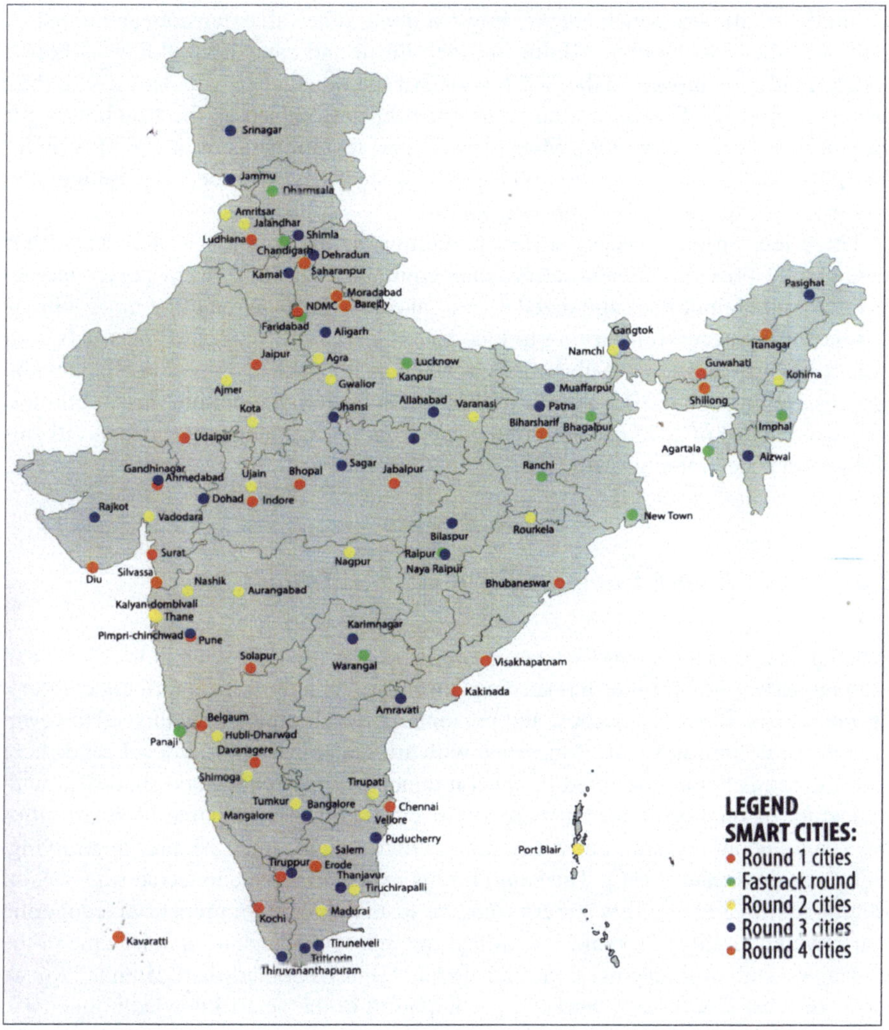


**Fig. 2** Hundred smart cities. *Source* Coffeetable Book Smart Cities Mission Journey so far, MoHUA, Jan 2019

cities in India, therefore, need to carry forward the motivations of well-being, equity, efficiency, and foresight and break them down into tangible indicators for enhancing the city's performance, while becoming digitally proficient. Here, institutional capacity must be enhanced through the use and education in technology, which shall ultimately lead to good governance and therefore sustainable urban development.

The projects strategized are in the form of four main categories namely Retrofitting, Redevelopment, Greenfield development and finally Pan-City Initiatives. These projects are categorized on the basis of the functions that they will serve within the smart city, which are city improvement, renewal, extension, and transformation of the existing city. Administrative reform and capacity building, although not explicitly mentioned in the SCM, is one of the crucial aspects. Along with that, the sustainability of resources has also not been introduced as the benchmark for the smartness of development. Most of the Smart Components of the SCM is therefore fairly rudimentary in nature and require a smartness of process, planning, and proposals for the city to become truly smart.

There are various success stories pertaining to the mission—Ahmedabad has been able to provide efficient, affordable, equitable and customized governance to its citizens. Udaipur has capitalized on its Lakeside Heritage to attract more tourists and has been successfully trying to face water scarcity through the construction of water pipelines and so on. Solapur has gone clean and has also become efficient and progressive through the development of physical capital that pollutes less. Ludhiana has set a time bar of 2025 and assures to provide enhanced quality of life to its citizens along with accelerated growth and sustained prosperity.

## 2.2 *Smart Living Components for Smart Cities*

When it comes to the smart living concept, there is no distinct scale in which we can limit ourselves. Smart living has varying definitions for architects, interior decorators, psychologists, lifestyle coaches, and personality development specialists. However, the kind of definition we are concerned with are scalable to the city level, since here smart living is being discussed in concomitance to smart cities, and shall deal with indicators of quality of life in these smart cities. The components of smart cities are a smart economy, smart mobility, smart environment, smart people, smart living, and smart governance [16]. The smart city is expected to be connected (in terms of smart logistics and mobility patterns that are sustainable), entrepreneurial (economic viability and vitality), liveable and pioneering in terms of social capital. Some of the indicators and components identified under "Smart People/Smart Human" are in terms of education levels, worker's participation in the smart knowledge-intensive sectors (R&D), voter turnout at different levels of elections, gender ratios, number of representatives per 1000 residents, language skills, participation in life-long learning and technological skills. Under the "Smart Living Category", the components include participation and funding for research, land availability for green and recreational spaces, public libraries, cultural venues, government spending on smart health care services for the public, tourism services, mass media access, adoption of international standards for enterprises and participation of people in industry-based trainings.

Indicators and components such as $CO_2$ reduction strategies, city and building-level energy efficiency, water efficiency, management of urban sprawl and peri-urban environmental performance, pollution control, participatory environmental decision making, the share of renewable energy, rate of recycled waste and growth of new building make up the "Smart Environment" category. It is these categories of smart human capital, smart living, and smart environment, together in combination with aspects like e-governance, entrepreneurship, and ICT that together constitute the umbrella term "smart living for smart cities".

A major component of smart living in smart cities is the component of sustainability. In the case of smart cities, technological advancement goes hand in hand with the sustainable development of environments and economies. Sustainability is the relationship between the elements of the economy, ecology, and equity [3]. Within the concept of sustainable development are embedded in the improvement of quality of life, respect for environmental limitations and investment in technological progress (Fig. 2). This is how smart cities and smart sustainable living are interconnected. The concept of sustainable living is best exercised through the concept of smart cities because cities represent the major causes of production, consumption and using up of our limited resources. The core components of smart sustainable living in a smart city can be visualized through smart transportation systems, smart government systems, smart educational systems, and smart health systems. Batagan also outlines the different processes through which these fundamental components of smart living might be affected. These include:

- Intelligent Educational Systems: data systems for key performance indicators; education clusters for stakeholder participation; cloud computing;
- Intelligent Transportation Systems: Intelligent traffic management and fees collection systems;
- Intelligent Health Systems: Integrated intelligent medical record databases connecting doctors, patients, and insurance companies; electronic medical services
- Intelligent Government Systems: Data processing systems; e-governance with financial and paper-work services; electronic dissemination of information to all citizens and participatory collaboration mechanisms.

**Smart Living Summarised**

Since smart living in smart cities is intelligent, wired, digital and inclusive, the concept of mobility is of utmost significance. Smart mobility within a smart city is not just limited to traffic and transportation management but represents an intersection of the different components of a smart city, i.e. A digital city, green city and knowledge city [5]. The link between smart mobility and smart living is not just limited to ICT enabled networks and improvements in flow of people and ideas across space, but also includes concepts such as safety, environmental protection, and individual well-being. Smart mobility impacts smart living by reducing greenhouse and other emissions, enhancing personal safety and reducing the travel time to generate a more holistic work-life balance. The interesting thing about smart mobility is that it tries to bring together the different processes which generate a smart city, i.e. public behaviour and technological advancement.

In the following chapter, we will be looking specifically in the domains of ICT based interconnectivity and communication, sustainability and smart mobility as components of smart living within the New Delhi Municipal Council area, the smart city of Delhi. The upcoming section presents the case study city followed by a section on examining the geospatial tools and techniques through which the networks and environments within a smart city are modified and how geospatial tools can enhance the smart living experience of smart cities.

## 2.3 *Smart Geospatial Living*

According to IBM, cities are products of a long term process of fast and ever-evolving changes which constantly reinvent the ways in which cities are functioning [14]. The most vital component of such ever changing cities is the way data is obtained, analyzed and employed in the improvement of people's lives and institutional capacities. The Massachusetts Institute of Technology observes urban areas to be "systems of systems" which thrive on the potential for optimizing their "digital nervous systems" to the sustainable, livable and equitable development of buildings, cities and finally regions [19]. Cities are also envisioned to be fields of continuous cross communication, feed-forwards and feed-back processes between these different digital nervous systems which include the networks and components referenced in the Fig. 3. Sustainability and efficiency, which is the ultimate goal of smart cities is realized through efforts to coordinate the spatial and functional interrelationships between the different parts of a city's systems. These are studied and intervened into through spatial information of maps, data snapshots, models, supply chains, different assembly and fabrication processes, motion and activities which are divided into synchronic and diachronic aspects. Smart cities involve multiple spatio-temporal scales and diverse components which can be handled most effectively through geo-spatial technologies. Although there is a lack of clarity about the core technologies for developing smart cities, it is undeniable that location information is a crucial element of decision making, and is identified as one of the core ingredients in making cities smart [8]. Locational data plays the critical role of driving technology deployment towards the goal of systematic integration, planning and management of urban systems with the help of geo-spatial tools.

Smart living in smart cities has economic, environmental and social components. Smart cities have the common goal of enhancing the quality of service delivery to its citizens through participation, open governance, open access to data and its widespread public dissemination. Here geo-spatial tools play an important role in enhancing communication and interaction between the administration and the people and drive development towards a more effective and sustainable future. In smart cities, understanding the needs of the people, at the right time also involves a locational component. This information is not just a point in space, but a three-dimensional area which addresses the surrounding environment and its context, thereby bringing a new perspective to the flows and causes of different problems.

**Fig. 3** IBM concept for smarter cities. *Source* https://www.ibm.com/smarterplanet/us/en/smarter_cities/overview/

> Situational Information including geographic characteristics (ex. Topography, arid areas …) and spatial understanding capabilities participate significantly in the design and operation of Smart City service and management infrastructures. [8]

Geospatial information enhances decision making by answering several important questions which have a locational significance in terms of events, thematic contents, statistics, data flow, routing, spatial content and historical context at a particular location. It also enables decision makers to articulate the impacts of different events and their feasibility through simulation and correlation of activities with respect to spatial contents of locations. Some of the significant ways in which geospatial tools are affecting smart living is noted below.

1. **Interconnected Smart Living Through Intelligent Utility Systems**: GPS (Global Positioning System), GIS (Geographic Information System), GLONASS (Global Navigation Satellite System) and Remote Sensing can be used as Geo-ICT can be used in creating new utility management systems that overcome the problems of locating lines and nodes in existing utility networks, inefficient network management, planning and maintenance [1]. GIS is also able to generate and surmise data of locations and movements of assets over time at different

geographical locations and predict accurate futures through real-time analysis and visualization of patterns of asset use [28]. Geospatial tools enable efficient utility management through [18]:

(a) Accurate locational representation of physical components of utility networks spanning all lines, devices and structures.
(b) Modelling of utility networks to demonstrate the activities and processes through which it functions. This done using typologies, flow magnitudes, switching, valving and network engineering analysis.
(c) Integration of the spatial database with non-spatial entities such as work management systems, accounting or consumer networks. Geospatial tools also help in linking locational characteristics of different networks like customer files and billing information with geodemographics. This enhances the ease of business and creates greater transparency and accountability to the public (Fig. 4).

Geospatial tools which can be used for this purpose are the following:

(a) Data Capture through base-mapping using data from utility agencies, commercial street files, raster backdrop from existing maps, vectors digitized from detailed facility maps and field surveys [18].
(b) Map creation using four main components: map sheet and editor, non-spatial data entry tables, field design component and a user interface. The points that need to be taken care of at this stage are topological connectivity and referential integrity and availability of ad hoc mapping entities.
(c) Facility management Applications using query/display, design order processing, maintenance and operation, analysis of engineered networks and consumer support systems, their monitoring and evaluation.
(d) Modelling Applications: For future planning, trouble/anomaly management and understanding the impacts of networks on the city and its environment.

2. **Sustainable Smart Living Using Geospatial Tools**: Sustainable smart living can take a variety of forms ranging from the inclusion of urban planning to building resilient communities and leveraging microfinance and access to health. Remote Sensing and multispectral data acquisition and geographical information systems are some of the most basic geospatial tools employed in the domain. Examples of GIS application in sustainable development include the following areas:

(a) Spatial Database Management using Clustering Techniques to study demography, infrastructure and land information. Data has a textual and a graphical attribute as well as a spatial representation and finds applications in marketing (large customer databases), city planning (housing, land values, real estate), environmental studies (earthquake, disaster management and resilience, forest resources, water resources and coastal ecosystem management) [17].
(b) GIS based Scenario building and data visualization to create equitable future plans for informal settlements. The cities can be analysed through the built

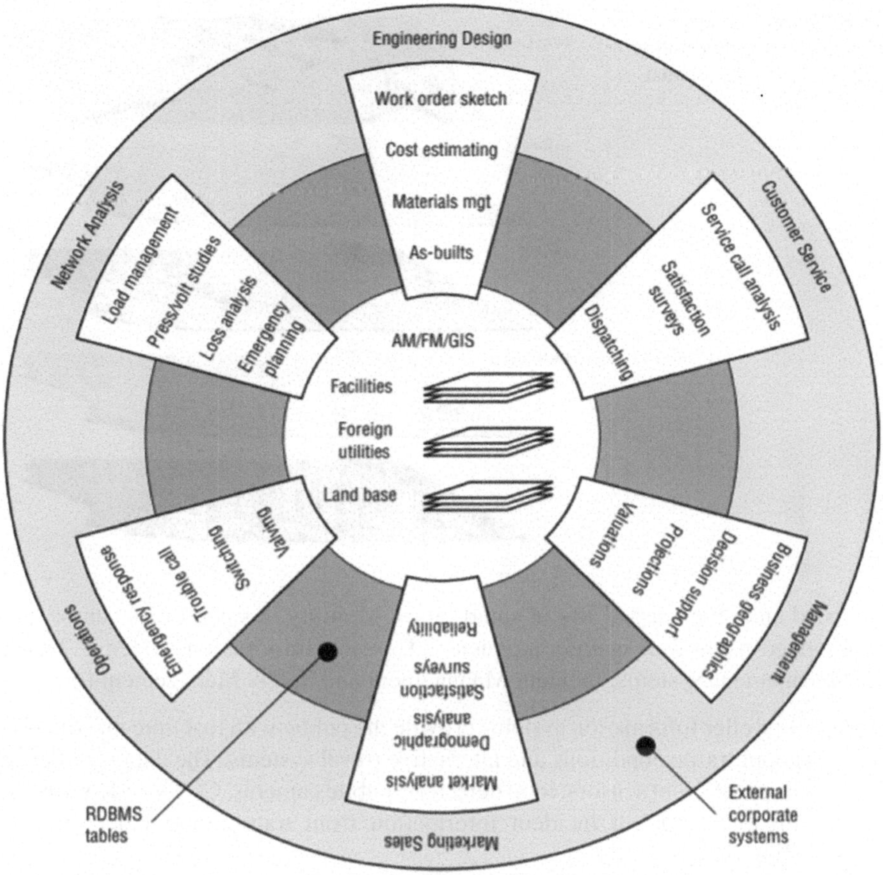


**Fig. 4** Geospatial intelligence for utilities management. *Source* https://www.geos.ed.ac.uk/

environment and the status of infrastructure networks using 3D Analyst GIS software and sustainable futures were designed accordingly [10].

In Ecological and Environmental planning GIS and Geospatial tools play the following important roles [20]:

(i) Identification of Analysis Criteria
(ii) Delineation of planning regions and identification of data needs
(iii) Building of basemap
(iv) Thematic layering over basemap using data acquired through automated and physically accumulated sources (Fig. 5)
(v) Creation of overlays and models
(vi) Refinement of models through evaluation of results.

3. **Fast and Mobile Smart Living Through Geospatial Technology**: Transportation infrastructures are one of the most significant initiatives to transform the

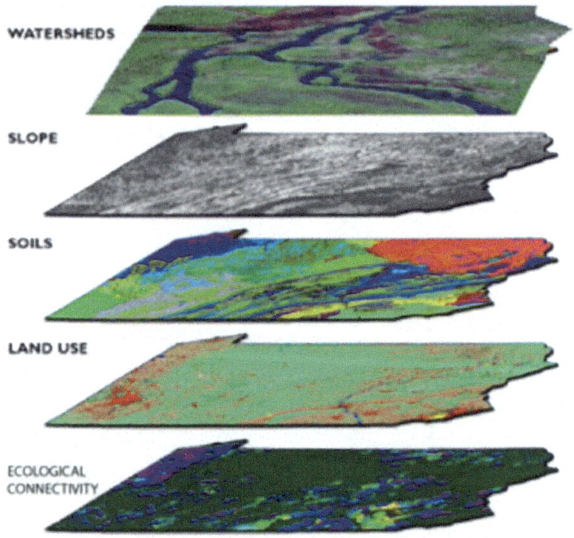


**Fig. 5** Geospatial layers for environmental sustainability analysis. *Source* Geospatial World (2012)

speed and efficiency of life in smart cities. Mobility in smart cities can be categorized using four main components: Traveller Information Systems, Traffic Management Systems, Incident Management and Safety Management [6]

(a) Traveller Information systems provide the public with real time information about traffic conditions and alternative travel systems. The data is obtained from different sensors, loop detectors, public cameras, CCTV video streaming Sand current incident information from transportation departments (Fig. 6).
(b) Transcore, Kapsch, Iteris and Siemen's Statos represent some of the leading traffic management systems which integrate data of traffic flow along with efforts of planning agencies to design better and manage roads and road infrastructure in a more efficient and sustainable manner.
(c) Incident Management Systems identify incident locations and attribute them to their responsible agencies through a centralized GIS network which enables emergency responses as well as predictive traffic management.
(d) Greater safety on faster and more efficient road networks is envisaged through Safety Management Systems. This uses Advanced Driver Assistance Systems (ASAS) with inbuilt GPS systems to track and avoid collisions, unwarranted lane departures, irrational speeds and pedestrian attacks.

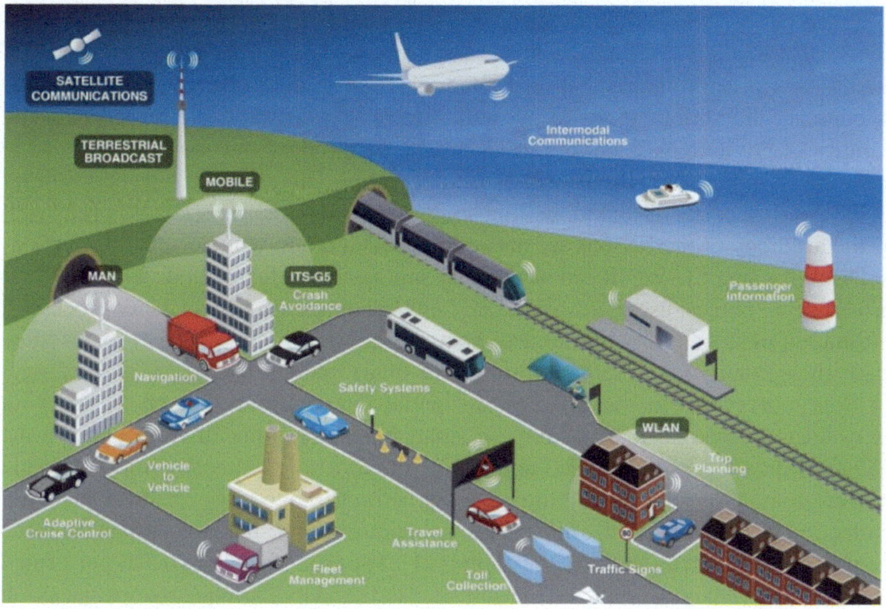


**Fig. 6** Smart mobility for smart living [9]

## 2.4 *Criteria for Evaluation of Geospatial Intelligence for Smart Living*

The evaluation of smart living in NDMC Area will be subjected to a three-tier process consisting of process, resources and finally the outcomes. For selection of smart living evaluation criteria no preceding literature that was complete and standardized was discovered and therefore a melange of suitable indicators from different sources have been combined. For evaluation of the project processes two important criteria, i.e. citizen engagement and use of innovative technology will be measured [4]. Here the process of citizen engagement will be evaluated in terms of ICT and geospatial approach, which the technology used shall be characterized in terms of sophistication of technology, use of embedded integrated systems and their respective techniques, nature of cyber security and data collection and innovative solutions to old and new problems. Resource evaluation will be done in terms of the type and nature of geospatial tools that were necessary and were undertaken. The resource evaluation will be done only in terms of technological utilization and not in terms of financial evaluation because that would exceed the scope of this chapter. We will therefore be looking at the integration of humanware, dataware and orgware that define a network of geospatial systems in smart cities [2]. Outcome will be measured in terms of overall output and nature of transformation in the quality of life of citizens and the impact of these projects on the nature of interconnected, sustainable and mobile smart living in smart cities.

Since the NDMC Projects will be evaluated on their nature of geospatial interventions, it might be useful to look at criteria for evaluation of geospatial methods of a cross-disciplinary nature [27]. In epidemiological studies, efforts have been made to compare the various kinds of geospatial tools that are available, in terms of the validity of predictions that were produced by the models and also in terms of the accuracy of input data with respect to ground reality. Therefore the type and validity of the data, the analysis and modelling strategies used in the geospatial tools as well as the accuracy of predictions are important consideration while evaluating geospatial tools for smart living. It can be done by summarizing the methods used and their prospective sources, use of covariates and interrelationships between city processes, estimation of uncertainty and unit of spatial analysis. It can be a part of resource analysis segment of the evaluation for urban smart living projects.

Data represents the backbone of geospatial interventions. Therefore, in trying to evaluate geospatial interventions in smart cities it is imperative to observe the nature of data that is being used in terms of the tools of data collection (automatic capture, remote sensing and satellite imagery, aerial photography or primary surveys), collection of spatial attributes and base map creation as per standards, urban database generation and its availability to the public [11]. The type of analysis that is performed both statistically and spatially on the collected data and its subsequent transformation into policy proposals also need to be monitored. It can be undertaken on scales ranging from the region to the building level and the best initiatives are also the most scalable ones. The scalable initiatives are generally those that resonate with the city's existing data collection and processing endeavours and fit into the larger ambit of urban planning in the city.

From the performance and outcome point of view smart living processes in smart cities must take into account efficiency of resource use and reducing environmental demand and drive the city towards a greener, more creative economic future. Smart living would also include a more holistic and democratic conversation between policy, policy makers, the government and the people and geospatial tools could affect more efficient and responsive conflict mitigation and incorporation of objections and suggestions into the project planning. ICT driven Smart Living processes need to be examined in terms of accessibility and use, health and wellbeing, appearance and aesthetic, management, environmental consciousness and safety and security [11]. Accessibility may include user-friendliness and ease of availability, provision for people with disabilities, sustainable liveability and mobility, flexibility and multifunctionality.

The multifaceted criteria for evaluation of geospatial smart living initiatives are mentioned above are tabulated in the form of table given in Table 1.

## 3 Case Study—Delhi, India

Delhi, the capital of India, has seen both glory and destruction in its long history. It has been plundered time and again only to bloom from its ruins to become the capital

**Table 1** Evaluation criteria: geospatial intelligence for smart living

| Criteria for evaluation | | Elements of smart living | | |
|---|---|---|---|---|
| | | Humans | Environment | Geospatial technology |
| Process evaluation | Citizen engagement | Mode of citizen engagement | Nature of environmental resources involved in the engagement process | Sophistication of ICT in geospatial citizen participation |
| | | Outreach and effectiveness | Environmental consciousness in management of citizen participation | Sources/companies involved in the geospatial components of citizen engagement |
| | | Appreciation for technological intervention | | Accountability and Transparency procedures adopted |
| | Use of ICT in projects | Categories of technical support teams (interdisciplinary nature of interventions) | Generation and handling of solid waste (including e-waste) during project implementation | Geospatial models/methods used |
| | | Orgware—organizations involvement and integration | Impact on pollution, environmental resource management, energy consumption quality of urban environment | Spatial and statistical analysis performed |
| | | | | Estimation of accuracy and uncertainty at every stage |
| Resource evaluation | Data | Involvement of citizens in data collection and processing | Inclusion of environmental concerns during data collection | Modes of data collection (automatic capture, satellite imagery, remote sensing and others) |
| | | | | Integration of spatial and non-spatial data |
| | | Granularity and invasiveness of collected data | Scaling of environmental data collection | Accuracy of Input Data |

(continued)

**Table 1** (continued)

| Criteria for evaluation | | Elements of smart living | | |
|---|---|---|---|---|
| | | Humans | Environment | Geospatial technology |
| | | Data privacy concerns | | Use of co-variates and interrelationship/linkage analysis |
| | Database | Public access and responsiveness of platform for real time edits, changes, suggestions, etc. | Modes of accessing the database (physically accessible, online, etc.) | Database design and generation techniques and efficiency, real-time data handling techniques and updating strategies |
| | Information systems | Integration of humanware with hardware, software and orgware | Linkages with existing environmental Information Systems | Tools for multihierarchical Integration between information systems, ease of vertical and horizontal transitions |
| | | Human inputs in decision making | Translation of environmental concerns into policy | Decision support systems and validity of predictions |
| | Planning and implementation | Compliance with Standards and Service Level Benchmarks/Issuing Authorities | | |
| Outcome evaluation | Overall outcomes | Project objectives versus project outcomes | | |
| | Interconnected smart living | Horizontal and vertical links between organizations and people | Connection between people and the environment | Relationship between environment, people and geospatial technology |

(continued)

**Table 1** (continued)

| Criteria for evaluation | | Elements of smart living | | |
|---|---|---|---|---|
| | | Humans | Environment | Geospatial technology |
| | | Conflict mitigation, grievance redressal or emergency response | Interlinking of ecological systems with utility networks in the city | Integration of geospatial technology in day-to-day activities |
| | | Social integration, social mixing and social inclusion | | Scalability and replicability of technology at different spatial levels |
| | Sustainable smart living | Innovation/knowledge based economic growth | Improved quality of life | Popularity in use of sustainable technologies and subsequent impact on the sustainable technology Industry |
| | | Impetus for sustainable development | | |
| | | Creative industries | Reducing environmental demand through resource efficiency | |
| | Mobile smart living | Accessibility and mobility of all social groups | Health and wellbeing | Use of latest technology (integrated transportation systems) |
| | | Infrastructure improvement and capacity building | Safety and security | Use of innovative/context-sensitive mobility solutions |

of many powerful dynasties. Yet, its resilient and enduring culture and heritage has withstood the test of time and the city continues to preserve the remains of its seven cities which flourished during 1100 A.D.–1947 A.D., with New Delhi being its eighth city. Delhi today is an amalgam of historical and modern, traditional and contemporary seven cities.

In 1947, post-independence, New Delhi was officially declared as the Capital of the country (Table 1). However it largely remained a Chief Commissioner. It was only in 1956 that it was converted into a Union Territory and gradually the Chief Commissioner was replaced by a Lt. Governor. The year 1991 saw the National Capital Territory Act (69th Constitutional Amendment Act) setting up a system of Diarchy under which the elected State Government was given wide ranging powers except law and order and land which remained under the control of the Union Government. The Act was actually implemented in 1993. Delhi government administers the whole of National Capital Territory of Delhi (NCT-D with area of 1483 $km^2$) comprising of 3 Municipal Corporations of Delhi (1297.29 $km^2$), New Delhi Municipal Council (42.74 $km^2$), Delhi Cantonment Board (DCB with an area of 42.97 $km^2$), 165 urban villages and 52 census towns. Fully developed Urban area is 525 $km^2$. Outer Delhi Area (Including Urban Ext.) is 961 $km^2$. Population of Delhi as per 2011 census is 200 Lakh and projected population in 2021 is 230 lakh. Municipal Corporation of Delhi (MCD) came into existence on the 7th April 1958 under an Act of Parliament. Prior to that Delhi Municipal Committee (DMC) was the principal civic body of Delhi. On 13 January 2012, the MCD was 'trifurcated', i.e. split into three smaller municipal corporations. The North Delhi Municipal Corporation and South Delhi Municipal Corporation each contain 104 municipal wards; the smaller East Delhi Municipal Corporation contains 64 wards.

Since nineties MoHUA has been implementing various schemes for the development of urban areas in India. During the first decade of twenty-first century schemes relevant for Delhi's development were BSUP, Infrastructure development of Satellite towns, Jawaharlal Nehru National Urban Renewal Mission (JNNURM), National Urban Livelihoods Mission (NULM) and lately Swachh Bharat Mission (SBM), Smart City Mission (SCM), Atal Mission for Rejuvenation and Urban Transformation (AMRUT), Pradhan Mantri Awas Yojana (PMAY) (Urban)—Housing for All. In all these schemes, MoHUA is advocating the use of the geospatial tool for during basic locational analysis and has made mandatory to have a GIS base map. For Delhi the creation of spatial infrastructure started with the commencement of Delhi State Spatial Data Infrastructure (DSSDI) project in 2008 which took 4 years to generate 184 layers of topographic data. Survey of India (SoI), on behalf of Delhi government, and with the full assistance of Navayuga Spatial Technologies executed the DSSDI to build up an integrated geospatial information system including 3D database for Delhi. The DSSDI project data used advanced geospatial tools to create a neighbourhood level granular database of 1:2000 scale. Before the project concluded the Delhi Geospatial Data Infrastructure Act, 2011 (DGDI Act 2011) was enacted with the spirit that every public service providing agency would use, contribute and share the Geo-Spatial data infrastructure to enhance the quality of their services in a cost-effective manner. Within the provision of Section 12 of the DGDI Act, 2011,

Geospatial Delhi Limited (GSDL) was constituted as special purpose vehicle (SPV) and was empowered as a company on August 2012 to take the custody of DSSDI data and to maintain and update geospatial data, to manage and share it with line departments of GNCT-D and other users.

Thus Geospatial Delhi Limited (GSDL), an enterprise of Government of National Capital Territory of Delhi (GoNCT-D) listed that 30 line departments which can utilize their GIS map and attached database for better planning and development. Of these about eighteen organizations and line departments of GoNCT-D are using the GIS base maps and GIS tools for better data analysis and coordination. These are Municipal Corporations of North Delhi, South Delhi and East Delhi; New Delhi Municipal Council; Delhi Development Authority, Delhi Metro Rail Corporation Ltd.; Health Department, Education Department, Social Department, Forest Department of GoNCT-D; Delhi Disaster Management Authority, Delhi Pollution Control Board, Delhi Fire Service, Delhi Integrated Multi-Modal Transit System Ltd., Delhi Transco Limited, Chief Electoral Office Delhi and Delhi Jal Board under GoNCT-D; Delhi Development Authority; few other portals hosted by GoNCT-D are Plan Dig Monitoring Portal, Ease of Doing Business (construction permits), and Citizen Geoportal.

Application of geospatial tool by the Department of Posts, Ministry of Communication, Government of India, for a Pan India initiative i.e. India Post Payments Bank (IPPB) portal which apart from providing services like—mobile banking, SMS banking, missed call banking, phone banking, QR Card, IPPB and Post Office Savings Account (POSA) linkage, also provides the option of locating branch, post office or Gramin Dak Sevak (GDS) across various states and UTs of India.

## *3.1 Smart City—New Delhi Municipal Council*

Lutyen's zone of Delhi, the historical power centre of India, is administered by a municipal council known as the New Delhi Municipal Council (NDMC). Its jurisdiction of 43.7 $km^2$. is known as the NDMC area which is the greenest part of urbanised Delhi as its green cover is 48% as compared to 21% for whole of Delhi. The NDMC is regulated by the Delhi Chief Minister. Its demographic status is tabulated in Fig. 7 (Table 2).

### 3.1.1 Evolution of New Delhi Municipal Council (NDMC)

In 1911 the British government decided Delhi to be India's capital and the Viceroy's residence and administrative setup moved from Calcutta to Delhi. After investigating many sites for construction of new capital Raisina Hill was chosen. Team of Edwin Lutyen and Herbert Baker conceived a plan for New Delhi with avenues dominated by the Viceroy Palace (now Rashtrapati Bhawan), Circular Pillar Palace (now the Parliament House), Supreme Court, Central Vista North and South Blocks and offices,

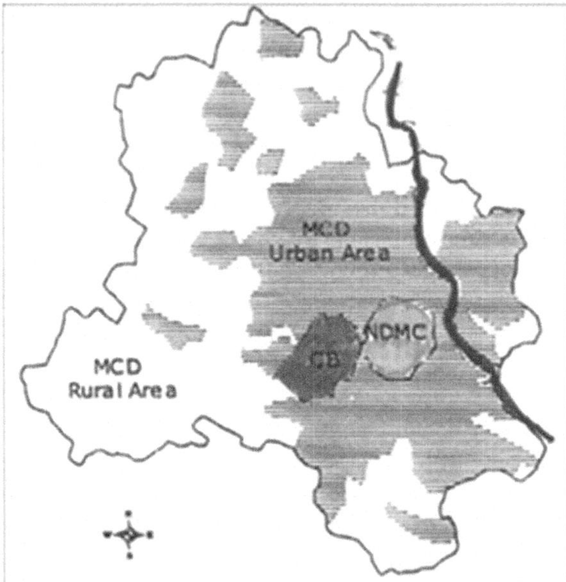


**Fig. 7** NDMC's demographic status. *Source* NDMC Developed for Round 1—Stage 2 of Smart City Challenge 2015

**Table 2** NDMC's Demographic Status

| Index | Facts |
|---|---|
| Area | 43.7 km$^2$ |
| Population (2011) | 2.57 Lakh (0.257 million) |
| Density of resident population | 7000 persons per km$^2$ |
| Green cover | 48% |
| Resident population | 3 Lakh (0.3 million) |
| Floating population during day time | 16 Lakh (1.6 million) |

*Source* NDMC Developed for Round 1—Stage 2 of Smart City Challenge 2015

as well as all the diplomatic missions that act as territorial bodies under the sovereign authority of their flag states, green spaces—parks and gardens. It was deemed necessary that plan's implementation would be transferred to a central authority instead of leaving the responsibility of construction and management to the local authority. This led to the formation of the Imperial Delhi Committee constitution on March 25, 1913. The Committee passed a resolution on 22 February 1927 for the adoption of the name "New Delhi" and the committee was designated as the "New Delhi Municipal Committee" which was approved on 16th March 1927 by the Chief Commissioner.

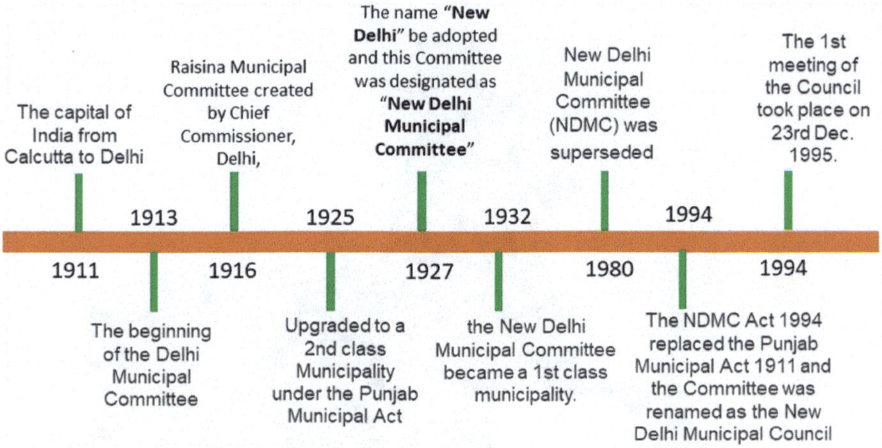


**Fig. 8** Evolution of New Delhi Municipal Council

The new capital officially inaugurated on February 15, 1931. Municipal Committee of New Delhi became a firstclass municipality in 1932. The New Delhi Municipal Committee (NDMC) ceased to exist in February 1980. The NDMC Act repealed the Municipal Act of Punjab, 1911, in May 1994 and the Committee was renamed the New Delhi Municipal Council. The Parliament passed the Act under Section 418 of the New Delhi Municipal Council (NDMC) Act 1994, the Central Government appointed a Special Officer until the new members were appointed. The Council's first meeting took place on 23rd December 1995. The Council continue for five years from the date appointed for its first meeting and no more, unless it has been dissolved under section-398 or any other statute for the time being in effect Section 5(1) of the NDMC Act-1994. Parliament enacted a Bill No. 55 of 1994 and the New Delhi Municipal Council Act. Following the approval of the Parliament Bill by the President of India, the Council was established on 10 November 1995. The Council's first meeting was held on 23 December 1995, convened by the Lt. Governor, Delhi (Delhi's Administrator) (Fig. 8).

### 3.1.2 NDMC—Governance and Administration

With the adoption of the Constitution (Seventy Fourth Amendment Act) some provisions in the Punjab Municipal Act 1911 had to be brought into line with Section IXA of the Constitution before 31 May 1994. With regard to the New Delhi Municipal Council, which consists of only 3% of the area and 3% of the Delhi National Capital Territory population, it was felt that a different type of legal system would be structured taking into account the special characteristics of the New Delhi Municipal Council area. India's government is almost the sole landowner and also owns about 80% of the buildings in the NDMC area. Efficient operation of municipal services

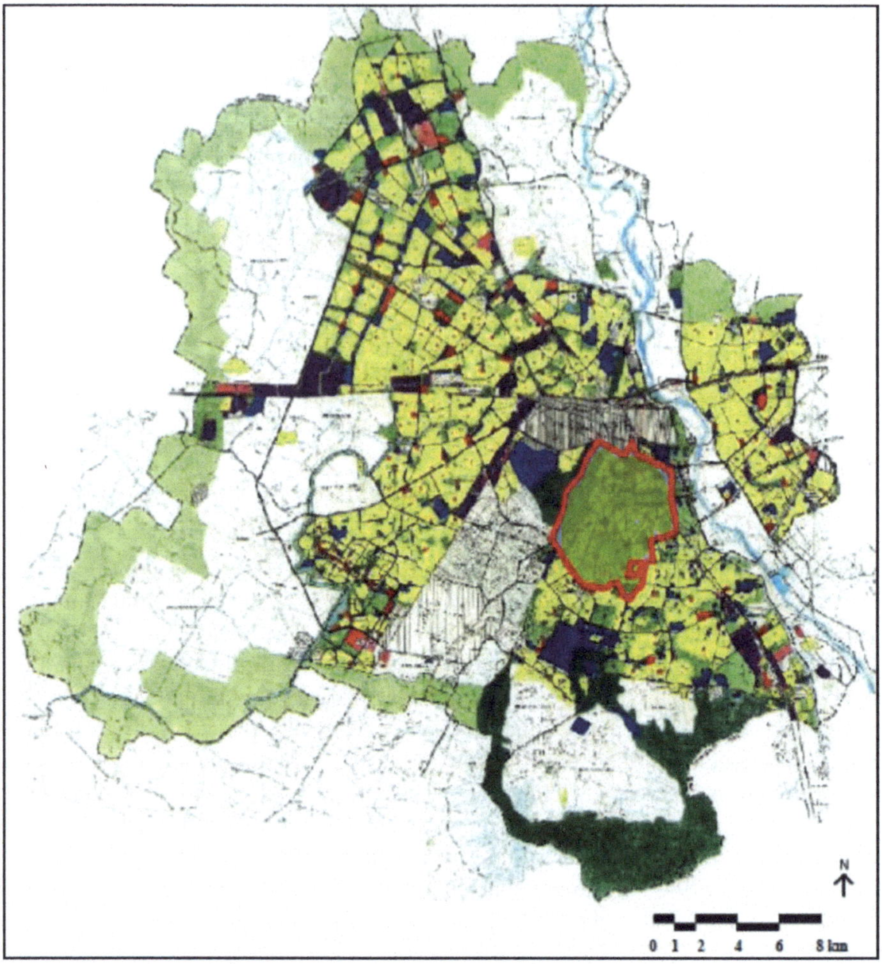


**Fig. 9** Location of NDMC Area w.r.t. Delhi (DDA)

in this region is essential to the country's internal image and is a factor that has a significant impact on the functioning of the machinery of government itself (Fig. 9).

## *3.2 Planning Framework*

New Delhi was designed in a geometric pattern over a triangular base, with three major functions at the apex viz. Commercial, governmental and recreational. Connaught Place, Government Complex and India Gate are the centers of these major activities within the NDMC area. NDMC area is on the south of Shahjahanabad,

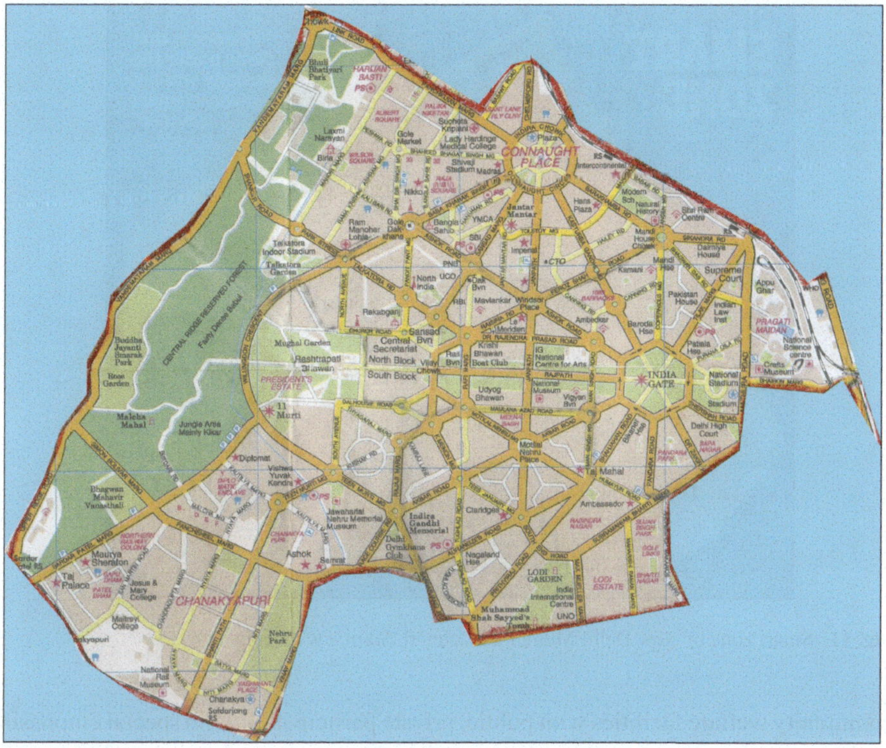


**Fig. 10** NDMC area road and localities map. *Source* Census 2011

Delhi's historic city and extends to the ring road. The areas, under the government complex, India Gate and the National Stadium, cover area of 22.3 km$^2$ and are known as the "Lutyens Bungalow Zone" (LBZ) (Fig. 10).

This area is included in Delhi's Planning Zone D as described in the Delhi Master Plan (MPD) 2021. The existing land use in Zone D indicates that more than 30% of the land is used for recreational purposes; 37% consists of residential areas (i.e. bungalows, government colonies and private colonies); 3% consists of commercial areas and the rest of the area consists of government, public and semi-public institutions and road infrastructure. State housing projects and bungalows make up more than 80% of the land within the metropolitan area.

## 3.3 *Vision and Objectives*

NDMC is committed to intensifying its efforts to provide citizens with better civic services and a multitude of visiting patrons from across the country and abroad. Aim of the NDMC is to improve the quality of life of residents by providing social and

**Fig. 11** Smart zone of New Delhi Municipal Council. *Source* www.ndmc.gov.in

community welfare facilities with public, private participation-with special emphasis on art and culture promotion, environmental improvement, cleaning and greening campaigns to serve as an "A Model Well-Planned Metropolitan City". NDMC Vision for Smart City is: "To be the Global Benchmark for a Capital City" (Fig. 11).

## 4 Smart Living in New Delhi

In New Delhi, the Smart Living Project is envisaged through the Smart City Programme, a flagship programme under the Government of India which started in 2015.

The New Delhi Municipal Corporation was a part of the Smart City Challenge, which was an effort in becoming one of the top 100 smart cities in the country. The framework for Smart City Development is given below in the figure. The reason for selecting the NDMC Area for the smart city project was that its land use is dominated by public and semi-public buildings whose redevelopment and retrofitting is comparatively simpler in comparison with commercial or residentially dominated areas.

The smart city projects in New Delhi are divided in the form of area based and pan-city projects. Although not mentioned overtly, the projects included could be defined as projects intended to make living in Delhi smarter. This chapter has made an attempt to categorize some of the smart city projects from the first phase of the Smart

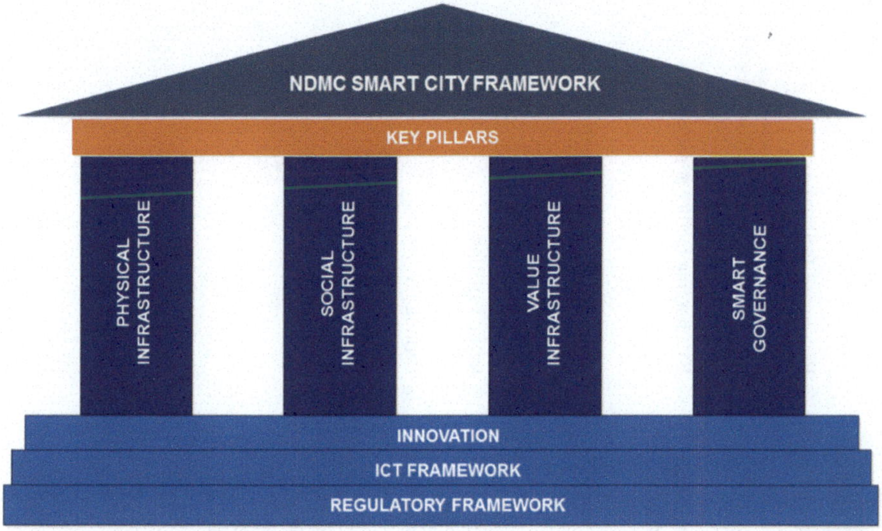


**Fig. 12** NDMC framework for the smart city project [21]

City Mission into smart living projects under the categories of interconnections, sustainability and mobility which are viewed as the core values of smart living in smart cities (Fig. 12). These projects can be viewed in the table described below Table 3. As per 2018, 60% of the projects are pending and many of the projects on ground or mentioned in the Smart City MIS are not in touch with the objectives and proposals of the Smart Cities Mission [4]. It lays down models of citizen engagement but does not talk much about slum area upgradation or social inclusion of diverse and disempowered communities in the city. The geospatial component within these smart living projects has not been clearly laid down leading to difficulties in their assessment and evaluation.

## *4.1 Evaluation of Geospatial Intelligence in the Smart Living*

The DSSDI database of GSDL is supposed to be the base layer for the mapping, analysis and execution of smart city Delhi. The projects show a considerable scope for inclusion of geospatial tools and intelligence for their planning, formulation as well as their implementation. However, the focus of these projects is not on geospatial tools and techniques. Besides slight mention of GIS based assessments, no concrete evidence of the process, data, analysis models have been revealed to the public. Therefore in order to obtain a comprehensive analysis of Delhi's spatial data infrastructure which forms the basis of smart living projects, DSSDI or Delhi's Spatial Data

**Table 3** Smart living projects under smart cities [22]

| | Smart interconnected living | Smart sustainable living | Smart mobile living |
|---|---|---|---|
| Area based development | Rooftop solar panels in public buildings up to 7.5 MW | Financial, identity, ticketing and access inclusion | Para-transit facilities including PELICAN crossing |
| | Rooftop solar panels in Pvt. Sector up to 7.5 MW | Happiness area for the cultural and social needs of citizen | Electric Vehicles (EV's) based last mile connectivity |
| | Transforming public toilets into smart public amenities centres | Transforming sub-ways into vibrant spaces ATM/pet adoption centre/Advt. etc. | EV charging facility at parking bay |
| | | Renovation of Gole Market, adding interactive Museum on History of Indian Civilization | App integrated cycle tracks |
| | Rooftop solar panels | Gateway to the world: on-street live video conferencing between people of Delhi and multiple Global Cities | e-surveillance including e-challan for traffic violation |
| | Augmenting existing mechanized road cleaning machines | Signature giant smart digital screen: traffic info/social messaging/alerts/cricket/advt | Parking for intermediate-public-transport |
| | | Transformation of electric-poles into smart poles with LEDs having incident-driven-controllers; communication-infrastructure, Wi-Fi access points, air-quality sensors, noise-pollution sensors | Smart bus stops |

(continued)

**Table 3** (continued)

| | Smart interconnected living | Smart sustainable living | Smart mobile living |
|---|---|---|---|
| | Automatic sewer cleaning machine | Hierarchical command and control centre | Sensor based smart parking |
| | 26 Green (horticulture waste) to gas Smart plant | Introducing signature initiative to the city's identity and culture | Pedestrianization of inner circle Connaught place |
| | | Happiness area | Multi-level automated parking at KG Marg, Shivaji terminal and near IOC building |
| | Geo tagging of bins | Sensor based common service utility duct | |
| | Providing new bins | Delhi international festival | |
| | | Behavioral transformation | |
| Pan city development | E-governance | Smart grid implementation | |
| | eLearning solution in all NDMC schools, virtual labs | 40 MW solar power projects | |
| | Centralized Student's health e-records | | |
| | Integrate all public medical facilities through Cloud-based e-healthcare system (Cloud provided by NIC under Digital India | Pan-smart water and waste-water management | |
| | Centralised Hospitalization facilitation for EWS for enabling them to access private hospital beds reserved for EWS, and provision of transportation service for transfer to private hospitals | Smart grid and energy management (ongoing) | |
| | Virtual medical service | | |

Infrastructure was analyzed in great depth and recommendations accordingly outlined. The evaluation also includes GIS based mapping project that was undertaken for the NDMC Area.

### 4.1.1 Development of Smart City Using Delhi State Spatial Data Infrastructure (DSSDI)

**Introduction**

The DSSDI's geospatial infrastructure is Delhi's basis for geospatially driven smart living infrastructure. It now spans the entire National Capital territory (NCT) of India and is the antecedent of earlier SDI (Spatial Data Infrastructure) projects in India [25]. The DSSDI project data uses advanced geospatial tools to create a neighbourhood level granular database of 1:2000 scale. For the upcoming Smart City Project, DSSDI under the Geospatial Delhi Limited (GSDL) provides a great platform for citizen engagement in e-governance. "Development of Smart City using DSSDI data" is an approved, World Bank funded project under the Department of Information Technology (DeitY), India for approximately 21.2 crores.

**Process Evaluation**

**Citizen Engagement**: The DSSDI was envisaged as a solution for monitoring of spatial datasets, developmental activities and to enhance citizen-centric services. Data of public interest in sectors such as health, education, etc. were posted publicly using the DSSDI infrastructure. Also mobile GIS application such as mHealth for assisting patients through emergency response was also being developed. It was proposed in 2014 that public geo-portal will be set up and crowd sourced data will be utilized for geo-tagging and updation of spatial and attribute data. The government to citizen component of e-governance is also to be strengthened through GIS based interventions for different kinds of taxation, etc.

Therefore the envisaged mode of citizen participation is not very clear through the DSSID's efforts, although future initiatives for citizen participation in the process of data collection has been outlined. It is therefore apparent that the popularity of geospatial data is an unexplored domain although large scale public information is available over online platforms. This also signals inefficiency and ineffectiveness when it comes to citizen outreach and generating a sense of awareness for spatial implications among the people. Since the mode of citizen participation is not clearly outlined, it is difficult to understand its implications for smart living. The potential for geospatial technologies to revolutionize citizen participation and promoting spatial literacy and appreciation among the people is exemplary. However, the agenda of DSSDI despite recognizing this, has a long way to go in actualizing its efforts.

As the model for citizen participation is presumably online, it has a substantial effect on smart living. Firstly, online citizen engagements, although vicarious, allow for a larger volume of information collection, storage and processing without the necessary resources and environmental constraints within which physical processes operate. Therefore the model for citizen participation, if online would also be environmentally conscious. Crowd sourcing, if combined with techniques for big data and social media analysis would massively transform the database and bring a qualitative dimension to the sensor and satellite imagery based information that is currently being used by the DSSDI. The Geospatial data Limited (GSDL), a Govt. of NCT of Delhi Company have been assisting the DSSDI in its venture, although the experts are definitely closely related to the field and bring in dedicated expertise to the spatial infrastructure. Since citizen engagement is only a prospect, its accountability and transparency measures have not yet been explored in depth.

**Use of ICT in Projects**: The NDMC Project also utilizes a geospatial map based portal with help from GSDL [12]. It is designed to act as a spatial decision support system has inbuilt functionalities for location identification, distance measurement, etc. It is built using the Arc GIS platform 10.0 with .NET technology. It maps the diverse land use zones of the NDMC and parcels of different commercial, residential, public/semi-public and other buildings and utility networks present in the zone.

At the orgware level it is a joint project between the GSDL and the New Delhi Municipal Corporation and build upon the premise underlined by the DSSDI. Therefore it is a collaboration between an local level administrative body and an expert technical support team. It appears to be a relatively new project undertaken by the GSDL, however it shows a good scope and potential to be replicated for the rest of the municipal bodies in Delhi and their subsequent integration into the larger unit as a whole. The DSSDI data which was a precursor to this project was developed using open source GIS standards and are shareable across all platforms. Most of the tools and techniques employed are in the form of digital inputs and display no harmful effects on the environment in terms of their procurement, use, maintenance or disposal. Therefore in terms of decision support mechanisms for smart living the use of ICT is sustainable and environmentally conscious, energy efficient and non-polluting infrastructure in nature.

The geospatial models used at this stage make use of the latest softwares in GIS mapping. They enable standard types of zooming, feature identification, feature creations, layer viewing, area measurement and map printing services for the user. It claims to include query based and analysis functionalities but it is not very apparent from the information provided. Also, it is claimed that much of the data that was collected has not been adequately compiled, making it difficult to include analysis based functions [12]. As a decision support system it does not clearly define its goals, objectives, nature of stakeholders or the kind of functionalities it can provide and is incomplete at best.

**Resource Evaluation**

**Data**: The data for Delhi's Spatial Infrastructure is obtained from photogrammetric capture i.e. aerial photographs and 3D, orthophoto and surveys for underground utilities leading to the formation of an land information system and the Urban Spatial Information System (USIS) for Delhi [25]. This data was mandated to be shared and accessed for 29 line departments and regulated for service level delivery under "The Delhi Geospatial Data Infrastructure (Management, Control, Administration, Security and Safety) Act, 2011". The data was collected at an extremely rich scale of 1:2000 which could be used for site-planning and neighbourhood design initiatives at the local level. Changes up to a level of 3–5 km could be monitored through 63 IP cameras and was used to detect unauthorized constructions and security events, collecting audio and video recordings at these instances. It acknowledges the possibility of data privacy risks due to the presence of personally identifiable location specific data and claims the use of secured encryptions to prevent data leaks and undue modification. However, privacy concerns related with the nature of data storage has been addressed and not the invasiveness of data collection procedures. Therefore, the data could be used for citizen surveillance without any democratic consent for the same.

Environmental data is supplied by the GSDL to Forestry, environmental departments as well as the Delhi Development Authority [24]. It has also conducted detailed riverbank mapping of the eco-sensitive O-Zone of Delhi for DDA's Master Planning. It contains a geospatial enabled disaster response center. Since most of the data under the DSSDI was obtained at the scale of 1:2000, it might be assumed that environmental concerns were also embedded within it. No particular mention of scaling and scalability environmental layers of DSSDI was found.

While trying to link attribute data with the captured spatial data, delays had occurred from the stakeholder department's side [25]. These delays were on account of different exigencies, manpower deficiencies and political events and could mean a certain dissociation between the time at which spatial data was collected and the time at which attribute data of the same location was obtained. This could also cause inaccuracies in the overall dataset, however no dedicated research has been directed towards the accuracy levels of the dataset. Also, the dataset does not take into account co-variates, or data whose variation depends on the other. Since co-variation is not recorded with clarity, it might be difficult to outline interconnections between the various layers and features on the existing maps. This will ultimately affect holistic, integrative planning action.

**Database**: The Geospatial Map based Portal (GMP) enables the citizens of National Capital Territory of Delhi to obtain spatial information on buildings, addresses, locations, measurement and suitable path analysis on the NDMC area [12]. However, as a database, it is not open to suggestions/changes/edits created by the people because of data integrity reasons. It has locked down the possibility of the public making changes to the existing database and has reduced public intervention to viewing only.

The database is based on a client-server architecture model and is available online. Other than the availability of internet speeds and the presence of the requisite hardware and software, no environmental concerns about the infrastructure were raised. The spatial database of the GMP only contains demography and utility spatial layers and does not include DSSDI's environmental data. The project was conducted using the latest tools and technologies provided by ESRI, which is a dedicated GIS based company. Data capture tools remain the basic spatial data collection methods, however no analysis is performed on the data, except the basic processing and map visualization. There is no provision for real-time data handling and despite talk of updation procedures, it is fairly unclear how updation is envisaged.

**Information Systems**: It is expressed that the DSSDI data is being utilized by all line departments, civic authorities, planning bodies and others as a standardized digital database. The DSSDI information system also aids the fields of city planning, urban development, land management, management of over ground/underground utilities, traffic management, property tax and revenue collection, updating property records, pollution and environmental monitoring, security planning and management and building construction change detection [25]. In order to bring about integration between the capacity of workers involved in the project and the end-users of the data (line departments), the GSDL held workshops about the updation and utilization of geospatial data maps. It continues to sensitize different sectors of geospatial endeavours through trainings and workshops. However the integration between different departments and a constant feedback for mutual learning of all stakeholder departments from each other's experiences is yet to be effected. Most of the decision making that is coming from the DSSDI's information is made through human endeavours and the role of GIS has been reduced to a spatial database only. There is vast scope of GIS for scientifically delineating solutions through data visualization, query based analysis and covariance detection is not fully realized by the users of the data.

The data is not linked to existing environmental information systems leading to the absence of a multi-hierarchical network of spatial levels at which environmental concerns could be analyzed. Most of the translation of environmental information into policy occurs at piecemeal levels, obtaining data from a plethora of sources like ENVIS, USIS, DSSDI and state level bodies without any vertical or horizontal integration between the departments. This leads to isolated decision making, wherein plans are not always sensitive to the contextual needs. DSSDI, although envisaged as a decision support mechanism is nothing but a simple database and most of the planning and decision making is done through human interventions, whose accuracy and validity trends are a grey area to analyse.

**Planning and Implementation**: The geospatial data infrastructure of the DSSDI was created under the "The Delhi Geospatial Data Infrastructure (Management, Control, Administration, Security and Safety) Act, 2011" and its legal entity was set up under section 25 of the Indian Companies Act (Geospatial Delhi Limited or GSDL). There were agreed service levels between the line departments and the establishment of regulatory authority for enforcing the Act's provisions. Delhi GeoPortal also enables query based access to multi-geospatial data and publish their own metadata on the portal. Smart phones using GPS and Web GIS were used for data collection in the field. There are links between the stakeholders and GSDL through SMS and email services. However no dedicated communication mechanism has been designed for location specific conflict resolution. The platform is complaint with OGC and ISO Standards and interacts well with external platforms. The infrastructure is made secure through layers of physical, software, administrative and access control security tools. Service level Agreements were made between GNCTD, Survey of India and Government of Delhi for building the data infrastructure and in the future citizen centric services will be delivered under multiple SLAs.

**Outcome Evaluation**

**Overall Outcomes**: The DSSDI tried to intervene into the absence of a standardized platform for GIS data. With that it also tried to bring about an awareness and capacity development among the departments who would be now using their geospatial data infrastructure. It also put an end to the repeated surveys that were to be conducted by each department and mitigated the absence of inaccurate, unreliable and redundant data. The primary objective of the DSSDI project was to build 3D spatial datasets using modern technologies and it achieved not just this, but a number of incidental benefits along the process as well.

The DSSDI was conceived with the idea that it would facilitate smooth and integrated planning of the NCT of Delhi and mitigate the planning chaos that characterizes its current landscape. Although it has complied with the standards and tried to meet most of its objectives, its primary vision for a holistic spatial planning of Delhi has not yet manifested. It is because of absence of coordination, vertical and horizontal links between the database and the government departments which should be constantly using it. Also much of the data collection strategies were viewed suspiciously by the local residents since the awareness campaigns of the DSSDI did not extend to citizen education [25].

**Impact on Interconnected Smart Living**: DSSDI's geospatial data infrastructure will enable both people and organizations to interact in a hassle-free environment. Better tools for data obtaining is a precondition for designing effective tools for citizen government interaction in e-governance. As of now, DSSDI's infrastructure is utilized for property tax collection and ICT enabled e-governance is enabling people to file their taxes online. Town down monitoring can also be conducted through the geospatial data infrastructure and unauthorized activities can be tackled accordingly. "Plan Dig Monitor" provides easy access to spatial and attribute data to the departments and single window interfaces help make application smooth and manageable. Mobile GIS helps connect citizens with local amenities and emergency services. This is a bold and vibrant step towards connecting the people with their policy makers and creating a scenario of interconnected smart living in New Delhi.

However, the DSSDI project is still at a nascent stage when it comes to adding a spatial dimension to grievance Redressal or participatory public action. It has not been able to create a public platform for data management and is not recognized by the public as a necessity in their lives. If the DSSDI project were to open a platform for land use and spatial information discussion through a participatory GIS venture, it would have created a geospatial contribution to a social conversation, bolstering a mix and inclusion of people in the decisions of their own city.

The DSSDI has 365 layers of information at a scale of 1:2000. It is has numerous layers on natural features and environmental conditions and can be clubbed with utility networks to create spatially responsive environmental plans. Therefore it has helped foster a positive relationship between people, their environment and technology. However, it geospatial information has become an important aspect of people's day to day activities, however DSSDI's data is yet to have a public outreach.

DSSDI has 10 identical monitoring centers which detect change and are embedded within a distributed, replicable and scalable architecture [25]. The storage facilities can cater to increase in loads as well as uneventful setbacks in the system, during which the load can be transmitted from monitoring center to another. For enhancing the scale of the project, upgrading the infrastructure is not a cumbersome process. Its scalability and Replicability are one of the major strengths of the DSSDI.

**Impact on Sustainable Smart Living**: The smart living benefits of the DSSDI project are sustained through the "Development of Smart City using DSDDI dataset" Project. Leased line departments can update the infrastructure in addition to control centers and the field surveys and department inputs by the GSDL. GSDL is a leader in the knowledge based economy sector and has liaisons with Geospatial Associates and Assistants for updation and creation of GIS data. GSDL has also created an impetus for standardized and quality data infrastructure and has led to departmental savings on an average of 30% due to availability of accurate and timely data for planning and implementation [25]. At the environmental level, it has reduced waste generation by the line departments through database visualization and eliminated the need for costly plotters and excessive paper-based print-outs. Also absence of invasive surveys and physical data collection has led to resource efficiency and reduced the pressure the departmental planning on the environment. The DSSDI has laid the framework for the city to use more sustainable means and methods for planning and implementation and has led to a boost in the geospatial entrepreneurship in the Delhi region.

**Impact on Mobile Smart Living**: The availability of geospatial infrastructure has freed up much of the time and resources needed for departmental work in the city and has made it faster for people to interact with the government, pay their taxes, file their complaints, etc. Availability of geospatial data has also enhanced the quality of infrastructure leading to the creation of better roads, smarter vehicles and more efficient, safe and secure pedestrian facilities. At the data infrastructure level, the GMP of the NDMC is visible to the public, but much of the DSSDI's database is not yet opened to the public. It will lay down the foundation for intelligent transportation systems and innovative mobility solutions for New Delhi, which are presently manifest through the bike sharing, Electronic vehicles and pedestrianization projects under the Smart City Project at NDMC. These projects are enhancing the accessibility and mobility of diverse social groups and disabled people and making roads and travel safer, faster and more effective.

## 5 Recommendations

1. The initiatives to promote citizen participation should adhere to a definitive and organised plan, which should be based on insights gained from geospatial data to ensure the plans are effective in realising their goals as well as efficiently doing so. Online citizen engagement, crowd-sourcing are some of the known techniques, already implemented by many private companies, in which organized data, in combination with statistical analysis would bring about a qualitative dimension to the information and enhance the quality of the plan implementation.
2. Current geospatial models do not have well defined targets, functionalities and properly organised data, though they make available lots of useful mapping and query based features. Fixing the lacunae will go a long way to ensuring the usefulness of leveraging ICT in projects. Since most of the technology is obtained from private players, the government users of geospatial data could ensure a research oriented atmosphere for collaboration with private companies as well as academic researchers in extending the cause of geospatial planning.
3. The data collection process should ensure consent among the stakeholders, and be robust enough to handle exigencies like unforeseen events which disrupt the collection process leading to gaps and inaccuracies in the collected data. Statistical analysis should be used to identify correlations.
4. A mechanism for modification of existing GMP of Delhi should be implemented, with each proposed change being scrutinized before adoption. In addition, environmental data needs to be incorporated into the GMP which is currently absent. Data updation and analysis beyond basic map visualisation should also be included in this portal.
5. Databases like DSSDI, ENVIS, USIS should be integrated with each other, and full scope of technologies like GIS like visualisation, delineation, query based analysis should be exploited to lead to integrated and context aware decision making.
6. Delhi GeoPortal should integrate a direct communication mechanism to cater to location specific conflict resolution.
7. Coordination between different government departments, databases should be ensured to realise the vision of a holistic spatial planning of Delhi. Awareness campaigns aimed at the residents of Delhi should be conducted to ensure public co-operation to the data collection process.
8. The DSSDI needs to create a public platform for data management, and ensure their data has a public outreach. It also needs to integrate a spatial dimension to grievance redressal and participatory public action. It should be able to reach a stage where the public can regard it as a necessity in their lives. It should be accessible to people in a language and visual ease which takes cognizance of their socio-cultural inclinations and should inspire the people to learn and understand more about the spatial components of their cities.
9. There should be an enhanced level of transparency regarding the sources, models and mechanisms for geospatial interventions in the Smart City Project in

Delhi. Public participation in project formulation and its subsequent geospatial component needs to be enhanced and imbibed in the public imagination. Spatial Planning should not limit itself to a city level agenda and should transform its relevance to a national level development agenda through vertical and horizontal integration across the nation.

## 6 Conclusion

The interface of smart living in New Delhi is realized through the Smart City Project. GSDL or Geospatial Delhi Limited is playing an important role in the conceptualization, development and implementation of spatial data services which are indispensible when it comes to smart city projects. For this reason, much of the DSSDI datasets and infrastructure is being modified and undergoing a process of augmentation and updation to assist in smart living projects. The Electric Vehicle Project, one of the most significant drivers of mobile smart living in New Delhi has included aspects of GIS assessment and evaluation in their framework. Bicycle routing, utility network upgradation and sustainable development initiatives, all require geospatial interventions in some capacity. In India, geospatial intelligence for smart living is a relatively new concept. It is in the nascent stage of its integration and assimilation into the public and policy imagination. Although much of the spatial data for NCT of Delhi has been accumulated under the DSSDI, the same cannot be testified for other cities in the country. In such cases the data infrastructure is so wanting that a complete conceptualization of informed decision making and planning infrastructure needs to be brought about.

Spatial data is not just about discrete elements which are visible on maps. It is intricately related with people who are the forces and drivers behind the manifestation of spatial forms and patterns. The current infrastructure tries to enhance institutional delivery, but fails to anticipate a realm for public engagement and people's participation in decision making. In fact, the geospatial infrastructures, if utilized to its fullest potential would create a vibrant series of democratic people's policies, reducing much of the paperwork, unaccountability, opaqueness and bureaucratic inefficiencies that often dissociate planning from its on-ground implementation. There is therefore a need for spatial literacy, beyond the purview of its use for little day-to-day activities, in the form of a dimension as important and portentous as time. Geospatial Intelligence is the culmination of all interactions between people, their environment and technology and it is to imbibe this sensitivity towards space and its implications that we need delve deeper into its potential for revolutionizing smart living in smart cities.

## References


1. Ajwaliya RJ, Patel S (2014) Design and development of GIS based utility management system at DOS Housing Colony, Vikramnagar, Ahmedabad. Int J Eng Res Appl 56–73
2. Banerjee A (2018) A geospatial information approach to planning for rurban clusters. Unpublished Bachelor's Thesis, 31 July 2018. Department of Physical Planning, School of Planning and Architecture, New Delhi, Delhi, India
3. Batagan L (2011) Smart cities and sustainability models. Inform Econ 15(3):80–88
4. Begwani C (2018) Evaluation of area based projects under smart cities mission—a case study of the NDMC smart city. Unpublished Bachelor's Thesis Report. Department of Physical Planning, School of Planning and Architecture, New Delhi, India
5. Benevolo C, Dameri RP, D'Auria B (nd) Smart mobility in smart city: action taxonomy, ICT intensity and public benefits. Draft Version
6. Bills TC (2013, July 161-4) Transportation: GIS for intelligent transportation systems
7. Center for Study of Science, Technology and Policy (CSTEP) (2015) Reconceptualising smart cities: A reference framework for India. Presentation, India International Center: https://niti.gov.in/writereaddata/files/document_publication/CSTEP%20Report%20Smart%20Cities%20Framework.pdf
8. Daniel S, Doran M-A.(2013) geoSmartCity: geomatics contribution to the Smart City. In: The proceedings of the 14th annual international conference on digital governance research, Quebec City, Quebec, Canada, pp 65–71
9. Dhade M (2017) Smart Mobility: A tool for smart and sustainable city. Retrieved from Geospatial World, 25 Mar 2017
10. ESRI (2011) GIS best practices: sustainable development. ESRI
11. Garau C, Pavan VM (2018) Evaluating urban quality: indicators and assessment tools for smart sustainable cities. Sustainability 10(575):1–18
12. Gupta AK, Kumar P, Sharma KP (2017) Development of geospatial map based portal for New Delhi Municipal Council. In: The international archives of the photogrammetry, remote sensing and spatial information sciences, ISPRS Geospatial Week, vol XLII-2/W7. Wuhan, China
13. Harrison C, Donnelly IA (2011) A theory of smart cities. In: Proceedings of the 55th Annual Meeting of the ISSS. Hull, UK. Vol 55:1
14. IBM (nd) Smarter cities. Retrieved 28 Dec 2019, from Smarter Planet: https://www.ibm.com/smarterplanet/us/en/smarter_cities/overview/
15. Khan S, Taraporevala P, Zerah M-H (2018) Defining Indian Smart Cities Mission Impossible. Econ Political Wkly 53:49
16. Lombardi P, Giordano S, Farouh HA (2011) An analytical network model for smart cities. In: Proceedings of the international symposium on the Anaytic hierarchy process, pp 1–6
17. Malik M, Singh DP, Sharma DA (2011) GIS and RS for sustainable development integrating with data clustering techniques. In: 5th IEEE international conference on advanced computing & communication technologies, pp 291–296
18. Meyers JR (nd) GIS in the utilities. Retrieved 2019, from GIS Book (Abridged): https://www.geos.ed.ac.uk/~gisteac/gis_book_abridged/files/ch57.pdf
19. Mitchell WJ (nd) Smart cities: group information. Retrieved 28 Dec 2019, from MIT Media Labs: https://smartcities.media.mit.edu/frameset.html
20. MJ N (2012) GIS-based environmental and ecological planning for sustainable development. Geospatial World, 16 Jan 2012
21. New Delhi Municipal Council (2015) Smart city proposal: developed for round 1—stage 2 of smart city challenge, Dec 2015. Retrieved 31 Dec 2019, from Mygov.in: https://blog.mygov.in/wp-content/uploads/2015/12/Smart-City-Delhi_v3.pdf
22. NIUA (nd) NDMC projects. Retrieved 25 Dec 2019, from smartnet.niua: https://smartnet.niua.org/sites/default/files/resources/NDMC_projects.pdf
23. OiER; UN Economic Commission for Europe; Graz. (2014) The smart city conference for sustainable cities. In: Conference review. Urban Future Global Conference, Graz, p 10

24. Prasad K, Vasanthakumar N, Singh S (2016) Delhi gears up to be a 'smart city', 3 Sept 2016. Retrieved 29 Dec 2019, from Geospatial World: https://www.geospatialworld.net/article/delhi-gears-up-to-be-a-smart-city/
25. PricewaterhouseCoopers (2014) Delhi state spatial data infrastructures (DSSDI). Case Study. Department of Information Technology, Government of NCT of Delhi, 6 May 2014
26. Probst L, Monfardini E, Frideres L, Cedola D, PwC (2014) Smart living: smart construction products and processes. Best Innovation Observatory, European Commission, European Union, Luxemborg
27. The Subnational Estimates Working Group of the HIV Modelling Consortium. Evaluation of geospatial methods to generate subnational HIV prevalence estimates for local level planning. Imperial College London, London
28. Tiwari A, Jain K (2014) GIS steering smart future for smart cities. Int J Sci Res Publ 4(8):1–6
29. Tranos E, Gertner D (2012) Smart networked cities? Innov Eur J Soc Sci Res 25(2):175–190

# Role of Smart Land Scape Architecture in Smart Development of the UAE

**Ashmita Karmakar and Ashikha Raoof**

**Abstract** Development of Cities are often judged by the economic condition, commercial activities and building forms. The emirates cities of UAE are known for high standard of lifestyle, safety, security and beauty. The city has expanded from a small trade center to a large commercial hub. The extreme weather of UAE gave rise to clusters of housing settlements with very less window openings that restricted direct sunlight and hot air within the habitable areas. Instead, light and air was allowed to pass through the small courtyards, corridors, sikkas, wind towers and vertical slits within the roofs of the buildings that could reduce the harshness of the climate and provide diffused light and cool air to the buildings. With the advent of technology and digitization, people's movement got restricted and the outdoor spaces and public realm started losing their importance. UAE Government is trying to conserve the courtyards, sikkas, walkways in old city and create new parks, playgrounds, event centers in the new area. The Government is forming guidelines to reserve a percentage of open space with community and connect these spaces with the pedestrian and bikeways within and outside the developments. The recent drive to integrate all communities with the help of public realm by using land scape features and allocating various active spaces is a move towards a unique healthy urban development. The research will prove how the weightage of Public realm is more than land value land value in sustainable urban design. This chapter will describe the typologies and hierarchy of open spaces and public realm of the UAE that have regained the importance under the governance of his highness and developed innovative ways to recreate and integrate with new activities supporting the urban fabric.

**Keywords** Urban development · Smart Land Scape · Infrastructure development · Sustainability

A. Karmakar (✉)
Khatib and Alami, Dubai, UAE
e-mail: ashmita.karmakar@gmail.com

A. Raoof
Middle East Investment and Management Consultancy, Dubai, UAE
e-mail: ashikha92@gmail.com



T. M. Vinod Kumar (ed.), *Smart Living for Smart Cities*,
Advances in 21st Century Human Settlements,
https://doi.org/10.1007/978-981-15-4603-7_5

## 1 Introduction

For almost two decades, UAE has proved to be a trendsetter in terms of embracing digital technology by adopting digital governance approaches, utilizing information and communication technologies for development of the region. The city is also adapting rapid societal changes and technological advancements by providing infrastructures to enhance businesses, adjusting its policies to give rise to ambitious endeavour and creating the "smartest" and the "happiest" cities in the world.

The region is also planning to expand its utilization of advanced digital means for sustainable growth and development with a better governance and public wellbeing as a part of the leadership approach and collaborative governance style.

The plan was that the smart city will nurture collaboration between the public and private sectors to achieve the targets in majorly six dimensions. Among the initiatives are [1]:

- Enhancing communication across the city through setting open data and shared data frameworks
- Creation of personal dashboards for the city's inhabitants and a "Dubai smart platform"
- Launch hundreds of integrated new services for mobility and transport
- Development of smart electric grid, among other initiatives contributing to environmental sustainability
- Creating smart district pilot projects
- Providing advanced analytical solutions for retail and trade industry.

Another initiative is in transforming and integrating municipal services that includes creation and connection of integrated solutions for public open spaces such as parks, beaches and public realm within and across the emirates as the major urban planning initiatives. This chapter will showcase such efforts of the municipalities within the emirates especially Dubai where smart solutions have created unique pilot projects that has become signature projects of the emirates.

## 2 Importance of Land Scape Architecture in an Urban Development

Seen in a long-term historical perspective, city space has always served three vital functions such as:

- A meeting place
- A marketplace
- A place for networking.

As a meeting place, the city provided opportunities for social exchange of information of all kinds. As a marketplace, the city facilitated commercial exchange of goods and services. And finally, public spaces enabled networking, access to and connections between all the functions of the city.

Within a span of only a few decades the city got transformed primarily to a working city and these public spaces for assembly got transformed into spaces of leisure and enjoyment and also a breathing space within the built concrete forest of the modern city development.

Over the period the public spaces, parks and playgrounds became an integral part of the city design and practice of landscape design became an important part of urban planning and design in the following ways [2]:

- Site evaluation: Landscape architecture became responsible for the evaluation of the site's characteristics and contamination level. The landscape professionals like engineers, ecologists, and landscape architects started site identification and site opportunity and constraints, and evaluate aspects of the landscape (geology, hydrology, scenic characteristics, etc.) for assessment and evaluation for safe construction.
- Preservation of the environment: As the global consciousness about the preservation of the environment is rising, the importance of landscape architecture is also going upwards very fast. There are plenty of landscape projects in the world that are all about environmental ecology. Besides bringing new solutions for all kinds of existing environmental problems, landscape architects also work to protect and preserve untouched natural treasures.
- Developing sustainable techniques that are environment friendly: This is the point where we need to discuss the word "sustainability". Sustainability represents holistic strategies which are making our environment healthier and cleaner. Landscape architecture has a major role in this practice, starting with energy saving and going all the way to water purification.
- Regeneration of polluted areas: Many regions of the world have been subjected to pollution, particularly from the manufacturing sectors. Although factories are crucial for our survival, we're all conscious of their negative impact on environmental health.
- Storm water management: Stepping further into sustainability, water management is a must-discuss subject. Degradation of ecosystems, climate changes and its relationship with water management is a real issue. Wetlands are endangered and are biologically very diverse. Purification of storm water, and flood control have become necessary for action as they have proved to endanger the mankind.
- Cleaning the soil and water: Due to increased soil contamination, the world had to come up with a cost-effective solution for its cleaning. Usage of plants as natural remediators has a major positive impact on the soil because they're capable of taking in the pollutants, concentrating them and metabolizing them in their tissues.
- Caring about plants' needs with a proper quality of soil, right amounts of water, sun exposure and good protection from pests in order to grow and give their maximum.
- Taking plantation to various vertical levels: As time passes by and the population rises, cities are getting bigger and bigger. Lack of space for gardens sounds

terrifying. Compensating the lost land on the ground and taking the greenery to vertical levels gives an answer to conservation of greenery. Other benefits of vertical greenery includes cleaning the city air, retaining rainwater, and increasing biodiversity.

## 3 Challenges in the Urban Development of UAE

With the growing development and influx of population, the consumption of resources is also rising. The UAE region is spending a substantial amount on the basic resources as the climate and the physical character of the land does not support the habitats. The challenges to the natural environment today are immense. The UAE faces the impact of rapid development and the effects of climate change and global warming. Protection of environment and sustainable development are key topics which receive great attention from the UAE Government. The major challenges faced by the UAE region that has become concern are as follows:

### *3.1 Climatic Conditions*

UAE's climatic conditions such as high temperatures, evaporation, relative humidity and low average rainfall, play a major role in the degradation of land and of fragile ecosystems characterised by vulnerable vegetation and erodible soil.

The human and climatic factors along with wind storms lead to soil erosion, which results in deteriorating vegetation and perhaps erosion and dunes formation. These dunes or quick sand may encroach to arable lands, civil facilities and public roads and sometimes cause damage to them.

Erosion is also caused by water due to surface water runoff or because of the raindrops impact on soil. The effect increases due to heavy rains and runoff and as water erodes and deposits soft soil components.

Canyons and valleys are formed and increase in size with time. Loss of soft clay particles and floods result in the prevalence of loose sand in the surface layer of soil which leads to formation of superficial loose sand layer exposed to erosion by wind.

Climatic changes and frequent drought periods have led to a decline and degradation in natural pastures, resulting in increased numbers of grazing animals per unit area, which in turn led to the disappearance of favourite grazing plants and the spread of least palatable plants.

## 3.2 Change in UAE Biodiversity

Huge urban landscaping and increasing population has resulted in the endangerment and extinction of many plant species in last few decades. The impact of this erosion in plant biodiversity has created environmental and socioeconomic problems which subsequently triggered the need of conservation of plant resources.

Drought is considered as the main threat to biodiversity along with coastal development, urbanization and overexploitation of natural resources (grazing, fishing, hunting, and water extraction).

Similarly, desert ecosystems in the western region of UAE has also been subjected to different types of human impacts (e.g. afforestation, grazing and urbanization), leading to the loss of biodiversity.

The plants species having coastal habitat are diminishing very fast due to rapid urbanization.

## 3.3 Lack of Natural Greenery

Most of the urban landscapes are inspired by western landscape designs primarily focusing on aesthetics. This artificial greenery has introduced many new species in the region. Green sector of UAE depends mostly on imported exotic species. Greenery at UAE comprised primarily of non-native species which are used mainly in the landscape. Non-native plant species need high maintenance and need substantial support systems to keep them alive. Moreover, these exotic plants are mostly introduced from the temperate and semi-temperate regions to the arid environment of UAE and have high irrigation requirements. The purpose of planting exotic species is to increase the vegetation area and produce more aesthetic and green environment in marginal lands such as those with salinity or highly drought prone areas. The exotic species may become invasive and create environmental risk.

Due to this reason its required to focus on the indigenous plants while selecting plants for landscapes. In UAE dates and Rhodes are two main agricultural plant species grown for fruit and forage purposes respectively. Alfalfa and Rhodes grass are mostly produced to fulfil the shortfall of feed from the rangeland for grazing animals. These introduced species consume huge quantity of sweet water, mainly from groundwater water source. Due to combined effects of lowering of groundwater levels and accumulation of salts at soil surface have decreased the productivity of land.

### 3.4 Extensive Water Consumption

UAE has limited fresh water resources with average annual rainfall ranges from 80 and 140 mm the groundwater potential of both deep and shallow aquifers is about 757.6 $km^3$ while fresh water is less than 7.5%. The annual groundwater extraction is about 2668 million $m^3$ compared to recharge i.e. 350 million $m^3$.

During the last three decades the excess pumping of groundwater for irrigation has decreased the groundwater flow by one tenth. Current extraction rates need to be reduced by at least 25% to achieve sustainable development and to mitigate the present severe groundwater depletion in the area. The main source of water supply for greening the desert is ground water which is 70% of the total water consumed in the country.

The green sector in the UAE consists of forestry, agriculture and the amenities. The forestry sector is used for the landscaping purposes and amenities which include parks, gardens and recreational areas while the agricultural sector is focused on crop production. Agriculture is the single largest consumer of water by sector, accounting for up to 95% of groundwater usage and total amenity water use which was estimated at 547 $Mm^3$/yr. in 2007.

The low water use of residential flats relative to villas and shabiyats are also due to the watering of gardens rather than the actual use by the residents. Agriculture in the UAE is living on borrowed time, including palm plantations.

In the UAE, landscaping sector faces challenges, in terms of irrigation resources. The potential concern related to the green sector in the UAE is that water requirement for green sector is increasing while the groundwater supply is vanishing. The difference between the (incoming) feeding and the (outgoing) consumption of water from the aquifer has resulted in dryness of wells, reduction in ground water recharge, groundwater intrusion with seawater and increase in the water salinity (MOEW 2015).

### 3.5 High Salinity in Ground Water

Globally 800 MHa of lands are affected by salinity, which represents 6% of the world's total land area (FAO 2000). Though salinity is a natural phenomenon however, cultivated agricultural land has become saline due to land clearing or irrigation and sea water intrusion. Globally 45 MHa land is salt-affected out of the current 230 MHa of irrigated land, which is equal to 20% of irrigated land (FAO 2000).

It is estimated that about 10 MHa of agricultural land is vanished annually due to salinization, of which about 1.5 MHa is irrigated land. Furthermore, the saline areas are increasing at a rate of 10% annually for various reasons. It has been estimated that by the year 2050 more than 50% of the arable land would be salinized Groundwater salinity, is one of the problem that sustainable development and greenery is facing.

The groundwater salinity in the northern part of the UAE has increased abruptly and so as the central part of UAE to some extent.

In UAE seawater intrusion is also a main cause of water salinity in the aquifer especially near the coast. Recently after the year 2000, sea water intrusion extended 8 km inland from the coast of the Gulf of Oman.

The above-mentioned challenges make the region arid and the opportunity of practicing landscape architecture very limited with very high maintenance.

# 4 Why Smart Land Scape Architecture Is Important?

Over the past decades, the landscape approach has been put forward as a possible decision support solution for several development issues on a landscape level. The landscape approach aims to integrate the objectives of different stakeholders in order to establish long-term sustainable growth. The pursued objectives are those of sustained economic and social development, combined with local biodiversity conservation. It is important to expand our knowledge and understanding of the success factors, barriers and stakeholders that influence inclusive and sustainable development on a landscape level.

Over time, the interaction between people and nature within landscapes has evolved under varied scale of landscape typology. With increasing globalisation, technological development and the integration of people in global supply chains, landscapes today increasingly are the spatial scale on which many different stakeholders, from global to local level are cooperating to balance their competing interests and risks in order to take the places to a different level.

Smart landscape architecture is a practice that enhances the greenery by taking into consideration for the conservation of resources. Smart landscape architecture is a practice that involve the human kind from all directions and of all ages to a common space that is used for meeting, leisure and entertainment.

it is a practice that enhances and beautifies an open space through smart features and minimum maintenance.

For places like UAE, smart land scape architecture is ornamenting the master communities and creating signature spaces that gives a special platform to the UAE development.

## *4.1 Lifeline for Modern Cities*

Green spaces are an excellent indicator of the ecological health of a city. The role of green spaces is vital for a modern city as green belts, natural reserves, urban parks, wetlands and green roofs have the ability to contribute positively to public health, youth development, community building and environmental conservation.

**Fig. 1** Views of Creek Park, Dubai

**Fig. 2** Views of Jumeirah Beach Park, Dubai

The green movement is sweeping across the world in one form or the other, and the Middle East is no exception. The Middle East is witnessing a steady increase in the number of protected natural lands, ecological reserves, wetlands and other green areas dedicated to providing healthy habitats for indigenous animals and plants.

Most of the neighborhoods in the Middle East have a park and there are plans to build more in the near future.

Creek Park, Jumeirah Beach Park, Mamzar Beach Park, are some examples of green oasis scattered across the concrete landscape (Figs. 1, 2 and 3).

## *4.2 Sustaining Water Scarcity and Enhancing Green Growth*

With scarce water resources, the city uses a huge amount of water in the development and maintenance of parks and other green spaces. Being the world's most water-scarce region, it would be illogical for Middle Eastern nations to divert highly precious natural water or desalinated water for urban landscaping activities.

**Fig. 3** Views of Mamzar Park, UAE

To maintain a fine balance between water scarcity and green growth, it is essential to implement water conservation strategies like greywater recycling, smart landscaping and modern irrigation techniques. There are various methods for reducing the water consumption. Dubai is using grey water from households, factories and using it extensively for all the soft landscape covers of the public realm and street side landscape areas.

### *4.3 Promoting Sustainable Landscaping*

The Middle East is characterized by harsh weather, degraded soils, and scarce water resources. However, careful planning and smart landscaping techniques has been able to reduce the consumption of water and creation and maintenance of green spaces [3].

Landscape design in the U.A.E. should consider natural resources, especially for the selection of the right plants, which should be adapted to the natural environment.

The shift to sustainable landscaping encourages recycling green waste for use as mulch, compost and soil amelioration. Composting turns green waste into a range of high quality, natural soil conditioning products and potting mixes used in landscaping, horticulture and agriculture. These products take green waste back to earth, enriching the soil and helping to grow greener pastures. Recyclable green waste includes grass clippings, weeds, hedge trimmings, branches, leaves, floral decorations etc.

A sustainable and green landscape is an important component in every city's architecture and urbanization as it contributes to a healthy living environment. Nowadays work is performed within parameters that are both cost effective as well as environmentally conscious. In quality, the expectation is based on reliability and world class standards. This has pushed landscape designers to raise the bar and focus on formulating inclusive designs adapted for the region, rather than focusing on superfluous and random designs that are at odds with the climate and culture of the region.

### 4.4 Expansion of Green with Low Water Consumption

In a bid to transform UAE, significant investments have been made in landscaping and horticultural projects. However, successful urban landscape planning does require vast amount of irrigation water daily to ensure that landscapes are built and maintained to perfection. This in turn involves heavy usage and wastage of water and many other precious resources. Another important part is the cost involved in landscaping, which is many times overlooked in the excitement of seeing things green [4, 5].

Currently landscape designs have evolved to include xeriscaping, water conservation and plant palettes, reduction in the use of lawn grass in areas. Xeriscaping is derived from the Greek word xerox, meaning dry. It refers to the creative use of landscaping and gardening to reduce or eliminate the need for supplemental water from irrigation. It is designed especially for areas that are susceptible to draught or properties that want to practice water conservation. Deciding the plant palette is an important component of xeriscaping. Native species are resilient, beautiful, and have reduced water needs. These plants are born to thrive on very little while other ornamental plants generally used in building grand landscapes are barely surviving in the arid condition of this region. They require much more water and maintenance, thus increasing the cost of the landscape plan.

Xeriscaping in the U.A.E. is focused on using aesthetically-appealing and drought-tolerant native plants instead of water-thirsty exotic plants [6].

"Native plants like Ghaf (Proposis Cineraria) and Acacia (Vachellia Nilotica) are known to help fight desertification and reduce sand movements, offering shelter to many native species", explains Tatiana. Ghaf is a drought-tolerant, evergreen tree which can be seen growing on low sand dunes, undulating sand sheets and along margins of gravel plains of Abu Dhabi, Dubai, Sharjah and Ras Al Khaimah [7] (Fig. 4).

Segregating "thirsty" plants from those that require less water, saves unnecessary watering for plants that do not require it. This will not only save water and money, but it can also save the plants, since many a plant is killed through over watering.

**Fig. 4** View of garden decorated with local plant species, UAE

Another common element in xeriscape landscaping is the reduction of lawn grass areas, as lawn grass is often one of the worst offenders against water conservation. It is necessary for landscape designers to reduce the use of lawn grass in areas that are not accessed for play or recreation and replacing it with the use of low growing native groundcover species that require less maintenance and use less water, but still provide a 'green look'.

Soil health is also a crucial factor for the protection of green spaces in desert-dominated Middle East. Sandy soils have larger particles and do not retain water well, but provide good aeration. Adding organic matter, such as compost, to sandy soil can improve its quality, enabling better water retention.

Greywater and sewage water recycling can be instrumental in easing water availability problems for the maintenance of green spaces. The use of greywater or Treated Sewage Effluent for irrigating landscapes is vital to sustain parks, golf courses, resorts, streetscapes etc. Treated sewage is actually a better alternative than desalinated water as plants need some of the minerals for growth that are removed from seawater during the desalination process [8, 9] (Fig. 5).

## 5 Practice of Smart Landscape Architecture

UAE has launched various initiatives to enhance the green open spaces, conserve the green species and expand the network of green open spaces and public realm with the help of public and private participation [4, 5].

Various technologies such as Information Communication Technologies [ICT] and the Internet of Things [IoT] into its hard and soft infrastructures are applied to strengthen the networks and services to make them more efficient, intelligent and collaborative ecosystem.

UAE is growing with an ambitious plan to become one of the most sustainable and smartest cities in the world.

UAE desires to be known for its landscaped gardens, parks, and lush green public realm in near future. To make the dream come true, UAE is immensely progressing to adapt the smart city initiatives with new generation of smart projects, making the country a testbed for innovative designers to push the boundaries within the built environment.

Conservation of water, use of renewable energy, use of sensory devices, paperless transactions, reduction of carbon footprint became the prime agenda for the future growth of the country.

The similar criterion is applicable for the growth and development of greenery within the city. Restriction of water resources, recycling of water, sensory devices to control the resources and use of smart structure to enhance the landscape architecture of the city has become a part of the vision of the sustainable city (Fig. 6).

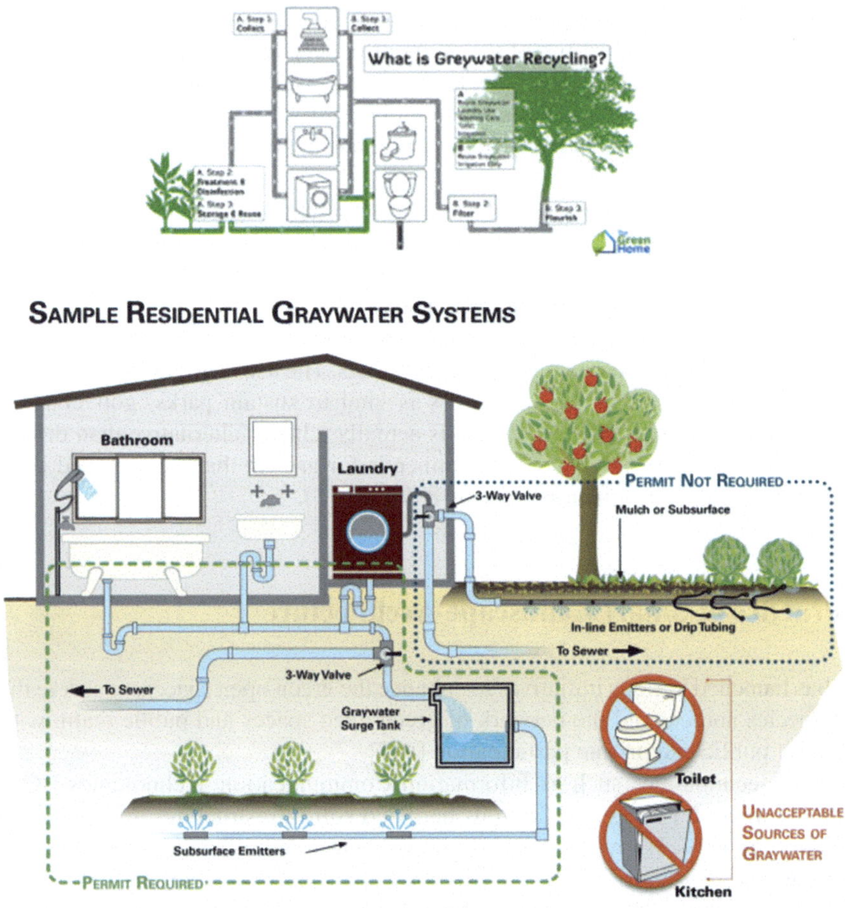


**Fig. 5** Figure showing the recycling process of gey water

## *5.1 Initiatives by UAE Government Through Guidelines, Applications in Approval Process of Master Plans*

### 5.1.1 Initiative by Capital of UAE—Abu Dhabi

The Capital city of UAE, Abu Dhabi has published a Public Realm Design Manual (PRDM) represents excellence in planning. The Manual identifies an Emirate-wide approach to public realm projects and looks at the social and contextual aspects of planning and design to create a high quality of life in Abu Dhabi. A collaborative initiative by the Abu Dhabi Urban Planning Council (UPC) with various stakeholders

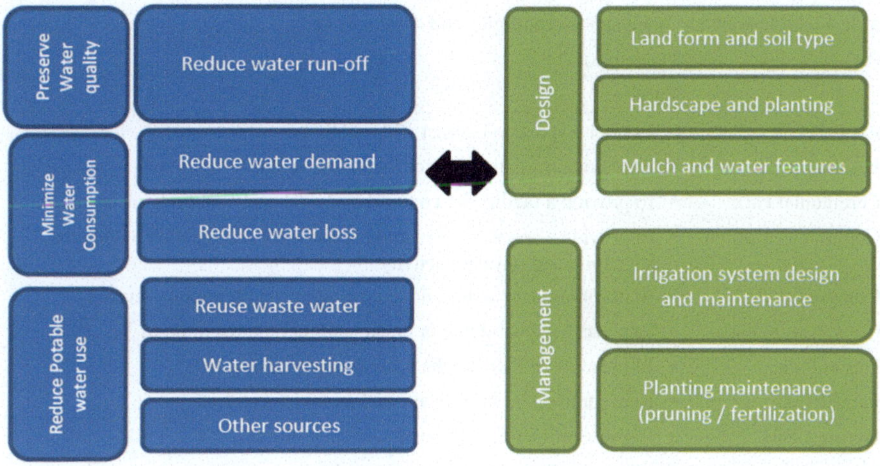


**Fig. 6** Figure showing the smart components of land scape architecture

has enabled to develop tools to build a public realm that work together towards achieving Vision 2030.

An extensive opinion survey was conducted to gather input on how people make use of parks, streetscape, waterfronts and public places and what they feel is currently missing in terms of programming and design to help in developing policies and design guidelines for the Public Realm Design Manual (PRDM).

The Manual aims to provide a comprehensively planned and well-designed public realm, which serves the needs of residents and visitors alike, providing a memorable experience by promoting quality public spaces and routes that are attractive, convenient and safe for all.

The Manual also reflects the enhancement priorities for the urban development projects to ensure safety and comfort, preserve heritage and culture, and improve accessibility to places of leisure while enhancing connectivity and mobility covering parks and open spaces including reserves of ecologically sensitive areas, national and regional parks, commemorative places of National interest and heritage, protected archaeological sites, parkways and boulevards, neighbourhood and urban parks, public beaches and coastal parks, fareej-scaled meeting and play spaces and shaded links between parks.

Some of the guidelines for the parks are as in the table

The Manual influences and informs the future design of all exterior places, linkages and built form elements such as streets, pedestrian ways, bikeways, bridges, plazas, nodes, squares, transportation hubs, gateways, parks, waterfronts, natural features, viewing corridors, landmarks and building interfaces.

By comparing the existing level of service in Abu Dhabi with that of other world-class cities along with Abu Dhabi's vision for the public realm, three key Level of Service (LoS) Standards were identified to create a world-class public realm. These are as follows:

**Table 1** Table showing open space typologies and details, UAE

| Park typology | Description<br>Purpose |
|---|---|
| Art Park | • To offer a space to interact with public art |
| Baraha | • To provide a small space for local passive use |
| Ceremonial Park | • To provide a facility for community-wide events and commemoration |
| Community Garden | • To provide a space for gardening activities, beautification. education, recreation and physical activity |
| Community Park | • To accommodate active and passive uses and community events |
| Conservation Park | • To preserve, rehabilitate or create natural features or areas |
| Desert Park | • To preserve or recreate the natural desert landscape |
| Family Park | • To accommodate local active and passive recreation |
| Heritage Park | • To preserve historic landmarks, sites or places |
| Linear Park | • To provide corridors for passive and active recreation |
| Meyadeen | • To provide a small central meeting area for passive use |
| Oasis Park | • To buffer the historic oases with a park for preservation and enhancement |
| Sports Park | • To accommodate sports activities and active uses |

- The provision of 1.3 hectares of developed parks per every 1,000 residents
- Maintain 20% of developed areas as open space
- Ensure the maximum distance to the nearest park is no more than 350 metres or a 10 min walk.

The above mentioned standards are to be used to quantify the amount of parks and/or open space that must be provided to meet a community's basic needs and expectations at present and in the future. Tying the Community Together Using public realm networks designed specifically for each focus area is aimed to create a unique development of community.

Each network illustrates the followings:

- Focusses on tying the community together with key connections and destinations
- Focuses on emphasising the Island's cosmopolitan waterfront characteristics by enhancing the urban grid and its connections to the waterfront
- Focuses on emphasising the Mainland's function as the Capital District by creating a system of linear corridors that link future destinations and major public spaces
- Focuses on maintaining the neighbourhood qualities of the City by creating a circular linked open space system with a primary and secondary loop to connect the urban areas to Al Ain's natural features, particularly the oases and wadis.
- Focusses on regulating the quality and appropriateness of public realm development, while also promoting a varied environment for a diverse public realm experience.

To identify the Level of Service, the public realm is broken down further into five hierarchy levels for each category of the public realm. The hierarchy levels are

**Fig. 7** Figures showing activities within UAE parks

Emirate, Municipality, City, District, and Neighbourhood. The hierarchy allows for a specific review of the Level of Service, clarifying the standards for a seamless public realm. The 27 public realm typologies create a unique character for each of them.

The implementation of the typology-based design guidelines enables the development of a unique public realm experience. Together, the methodology of making the public realm understandable using categories, hierarchies and typologies will help to implement the vision for a world-class public realm in line with Vision 2030.

To facilitate a distinct process for implementation, the PRDM includes a step-by-step implementation process. Each defined project begins with a policy review and an identification of how the project fits into the overall public realm system using the hierarchy, Level of Service and typology outlined in the PRDM before the project is submitted to the Abu Dhabi Urban Planning Council (UPC) for approval process.

As development occurs across the Emirate, each project is working towards a shared direction that brings various public realm elements together to form a complete public realm system [10].

### 5.1.2 Community Facility Guidelines—Dubai Municipality

Dubai Municipality has developed planning standards that aims to establish modern, sustainable, and integrated Communities fit with the values and the culture of Dubai & provide a distinctive Life style to the residents of Dubai.

These standards include the provision of community facilities within various levels—Neighbourhood, community, District and Sector further classified into low, medium and high density development.

**Fig. 8** Figure showing location of parks and open spaces AD and the linkages

A sample below is showing the various typologies of parks and green open spaces within each level of community development. The sizes of the open spaces provision are mandatory to follow for the approval of a master plan. Above all a minimum of 10% reservation of green open spaces/parks, green corridors etc. are mandatory within a masterplan community [11] (Table 3).

The planning standards/guidelines are developed by Dubai Municipality in cooperation with various governmental authorities and major developers, adopts a flexible, sustainable and competitive approach to Dubai's urban development—vision 2020 and beyond [12].

The key objectives include the efficient utilization of existing infrastructure, increasing investments opportunities, rationalizing a well-integrated landuse planning and mobility network. The plan covers the Dubai Emirates territories including offshore area within the 12 nautical miles, existing urban area, land committed for mega real estate projects, and desert and rural territory.

## 6 Making Landscape Smarter With Active Uses

As a part of UAE's vision for the future, an initiative has been taken by public and private stakeholders to activate the public spaces such as urban parks and gardens, streets, pedestrian walkways, alleys etc. with various features that will make the spaces more vibrant and inclusive in the urban development.

**Table 2** Table showing users, characteristics of open space distance, Catchment population, UAE

| Hierarchy | Users | Charactestics | Description features /Activities | Location | Service radius | Service population | Level of service (ha/1,000 population) |
|---|---|---|---|---|---|---|---|
| Emirate | Residents of the emirate | • Spacious land areas of Emirate-wide significance<br>• Natural, undeveloped landscapes<br>• Preservation of important natural features<br>• Emirate-wide important civic spaces | • Emirati significant public art<br>• Areas for emirate day celebrations<br>• Passive use | Dictated by natural Landscapes, major historical landmarks and monumental central spaces within developed areas | 100 km+ | Emirate | N/A |
| Municipality | Residents of a municipality | • Municipality-Important public spaces<br>• Historic and civic landmarks | • Major cultural events<br>• Historic and civic landmarks<br>• Monumental public art<br>• Generally passive use<br>• Suitable for Municipal-wide activities and gatherings | Dictated by major civic entitles, landmarks or historic landscapes of Municipality-wide significance | 25–150 km | Municipality | 0.2 |
| City | Residents of a city | • Important civic spaces<br>• Preservation of unique and historic assets<br>• Specialised destination sports facilities | • Public art<br>• Performance space<br>• Specialised sports facilities<br>• Active and passive use | Locations of major civic spaces should be central within densely populated areas. Additional City parks should be located around civic facilities and assets (museums, institutions, governmental buildings, historic landmarks. etc.). Larger City parks designed for specific uses are located adjacent and accessible to population centres | 2.5–25 km | 10,000–20,000 | 0.4 |

(continued)

**Table 2** (continued)

| Hierarchy | Users | Charactestics | Description features /Activities | Location | Service radius | Service population | Level of service (ha/1,000 population) |
|---|---|---|---|---|---|---|---|
| District | Residents of a District | • Serves multiple neighbourhoods<br>• Mix of daily use and important district-wide public functions | • Smaller scale public art<br>• Active and passive use<br>• Large athletic fields for organised and informal activities | Focused around developed population centres and can be co-located with other public uses such as schools | 0.3–2.5 km | 1.000–10.000 | 0.4 |
| Neighbourhood | Residents of a neighbourhood | • Integrated with daily lifestyles and activities<br>• Provide Neighbourhood users with a variety of play areas and gathering areas | • Features designated areas design for outdoor children's play<br>• Equipped playgrounds<br>• Smaller sport pitches/courts<br>• Abundant seating<br>• Abundant shade<br>• Active and passive use | Locations are within a maximum of 350 m of residents. Smaller typologies such as the baraha have a general distribution of 1 /developed residential block | 0.1–0.5 km | 150–1,000 | 0.3 |
| Total developed park level of service (ha/1.000 population) | | | | | | | 1.3 |

**Table 3** Table showing the community facility standards and details under DM

| Summary of community facilities standards for low density areas (<70 persons/Ha) | | | | | | | |
|---|---|---|---|---|---|---|---|
| Planning level | Type of facility | Pop. Served | Min. Site area ($m^2$) | Min. Site area per person ($m^2$) | Max. Diet to facility ($m^2$) | Building height | Total coverage area (%) |
| Neighborhood level | Local Mosque | 2000 | 1800 | 0.9 | 500 | Ground | 40 |
| | Retail Facilities* | 2000 | 400 | 0.2 | 400 | Ground | – |
| | Post Shelter | 2000 | 35 | – | 400 | Ground | – |
| | Local Plaza | 2000 | 1500 | 0.75 | 400 | NA | – |
| | Tol Lot** | 2000 | 1500 | 0.75 | 400 | Ground | – |
| | Neighborhood Park | 2000 | 4000 | 2 | 400 | NA | – |
| Community level | Juma Masjid | 6000 | 4500 | 0.75 | 1000 | G+Mez. | 40 |
| | Com. Shopping Center* | 6000 | 6000 | 1 | 800 | G+Mez. | 50 |
| | Children Nursery* | 6000 | 600 | 0.1 | 800 | Ground | 40 |
| | Kindergarten | 6000 | 3000 | 0.5 | 800 | Ground | 40 |
| | Primary School | 6000 | 8000 | 1.3 | 800 | G+1 | 40 |
| | Pvt General Clinic* | 6000 | 650 | 0.11 | Varies | G+1 | 55 |
| | Pvt Specially Clinic* | 6000 | 650 | 0.11 | Varies | G+1 | 55 |
| | Private Polyclinic* | 6000 | 650 | 0.11 | Varies | G+1 | 55 |
| | Playground** | 6000 | 3500 | 0.6 | 800 | Ground | – |
| | Community Park | 6000 | 12000 | 2 | 800 | NA | – |
| District level | Intermediate Sch. | 10000 | 9000 | 0.9 | 1500 | G+2 | 40 |
| | Secondary Sch. | 20000 | 12000 | 0.6 | 2500 | G+3 | 40 |
| | Pvt. Daycare Centre* | 12000 | 1200 | 0.08 | varies | G+1 | 55 |
| | Post Office* | 20000 | 2000 | 0.1 | 1800 | G+1 | 50 |
| | Health C. Centra | 30000 | 7500 | 0.25 | 2000 | G+1 | 50 |

(continued)

**Table 3** (continued)

| | | | | | | | |
|---|---|---|---|---|---|---|---|
| | District Park | 20000 | 20000 | 1 | 2000 | NA | – |
| Sector level | Civil Defense Centre | 50000 | 4000 | 0.08 | 4500 | G+1 | 50 |
| | Eld Prayer Area | 70000 | 14000 | 0.2 | Varies | Ground | – |
| | Public Library | 70000 | 6000 | 0.09 | 5000 | G+1 | 50 |
| | Police Station | 70000 | 5000 | 0.07 | 5000 | G+1 | 50 |
| | Elderly Rest House | 50000 | 2500 | 0.05 | Varies | Ground | 40 |
| | DM office | 50000 | 1500 | 0.03 | 4000 | G+1 | 50 |
| | DM Centre | 150000 | 4500 | 0.03 | Varies | G+1 | 50 |
| | Private Hospital | 35000 | 5000 | 0.14 | Varies | G+2 | 50 |
| | Gov. Ref. Hospital | 80000 | 12000 | 0.15 | Varies | G+2 | 50 |
| | Sector Park | 70000 | 500000 | 7 | 5000 | G for Park G+1 for facilities | – |

Given here are some examples set by UAE that have innovatively generated interesting activity spaces and made the open dead urban spaces come back to life.

## *6.1 Use of Bright Colours, Shapes and Materials as Street Furnitures and Urban Features*

One of the most attractive pop-up elements for children was definitely the implementation of bright colors in the otherwise gray space. And so is the use of different shapes in 2D and 3D land scape features. Use of various materials and colourful elements with a background of green space becomes the reason for stopping to sit down and take a break, instead of just passing through [13].

Landscape designers of Dubai have come up with focus on inherently sustainable practices, such as local material sourcing, reusing site-found materials and pioneering the use of native plants. In a Recently completed project in Dubai, a block was constructed over a seven-month period along the creek in Dubai Design District (d3) as a mural. Apart from this the urban park features recycled materials, countless play areas for children, an outdoor gym, sports facilities and food and beverage outlets (Fig. 9).

The parks act as canvases that come to life with people, artwork, events and performances. Many things from climbing walls to jump boxes in the gym, sloped

**Fig. 9** Figures showing the various colours in enhancing the open spaces/activity areas

surface of the block to create a slide in the place that makes landscapes unique and sustainable.

Throughout the park, nooks and crannies invite visitors to explore, play and rest with multiple play features such as volleyball nets, table tennis stations, an outdoor gym and an urban beach, a plaza, enclosed rock-climbing zone, skate bowl and basketball court and many more.

Spanning the 400 m site, there are green spaces with native plants populating the sandy patches and palm trees enhancing the land scape areas (Fig. 10).

**Fig. 10** Figures showing the various street furnitures enhancing the open spaces/activity areas

## 6.2 Use of Lighting to Create a Different Look

A Proper lighting can enhance the beauty of any landscape area. Without illumination, even a masterpiece can lose its splendor in the evening. To create an elegant look which can emphasize the shape and attractiveness of every particular piece, architectural and selection of proper decorative lighting are the perfect choice. Beautiful Garden is can result due to garden lights, that can also highlight special features such as conversation areas, walkways, fountains, and many more.

Lights have been used in various ways in various patterns, places to enhance the architecture and landscape features of Dubai and made every project in Dubai a special one.

In addition to fairy lights hanging from the tree branches, the tree canopy was lit from below with blue lighting, mix of warm and cool lights, use of colourful lights have always made innovative efforts in enhancing a space or a building during the dark hours.

Dubai has come up with a unique project—Dubai Garden Glow by using the lights in the most innovative manner. The park is created with the urban fabric of Dubai close to Zabeel park. The creation of the park is an effort to provide a place of entertainment, a breathing space within the urban fabric, a place for kids attraction and a place of landmark for the city of Dubai.

Dubai Garden Glow has been drawing visitors in droves since it opened. More than 10 million LED lights were used to make the garden glow. From the Ice Park, its brand-new attraction for the third season featuring ice sculptures, to the glow park with newly designed art installations, the garden has impressed UAE residents and tourists alike.

Every year as summer sets in, Dubai Garden Glow pulls down its installations to start afresh for the next season. In its second year, the focus was on erecting the dinosaur park with about 100 animatronic prehistoric creatures, in addition to a host of underwater—and forest-themed glow attractions (Fig. 11).

## 6.3 Use of Innovative Street Furniture

In the old days of UAE, the souks, sikkas and meydans of bastikiya used to be the outdoor majlis for the people. Streets had long benches, water fountains shaded with Areesh structure or a palm tree. Today's UAE has nodes are separated by over-scaled towers and wide motorways that alienate the streetscape.

Middle East is growing number of cultural projects and events. Yet street culture was a missing piece of puzzle in its urban fabric, an essential tool for establishing a solid cultural identity.

The vibrancy of street comes from one of the main components for solid street life is incorporating the right kind of street furniture. Today, street furniture has taken

**Fig. 11** Figures showing the lighting arrangements in Dubai Garden Glow

on a new identity as it shifts from regular wooden benches to interesting designs, quirky aesthetics and interactive features that are merged with city's design events.

The growing street culture can already be seen in many places of UAE bringing a more human level of interaction with outdoor cafes, seating, trees and walkways that could encourage a greater culture of lingering participation in the local scene.

There is also a strong potential for a new market and street furniture are of great demand. Designing street furniture will give local furniture designers a better command of their work with a creative selection of materials that withstand heat, direct sun, dust and humidity that are ideal for UAE's harsh climate.

## *6.4 Use of Digital Elements*

### 6.4.1 Palm Trees

Landscape with palm trees beautifies the space with natural look and also help to attract people and encourage them to come outdoors and use the public parks and open spaces with its character. Dubai is rolling out dozens of new Smart Palm stations that provide free wireless internet access in parks, beaches and other locations.

The UAE-made palm tree stations, standing six metres tall in the shape of a palm tree, also have fast mobile phone charging stations and screens displaying information on Dubai's attractions.

The first Smart Palm was installed at Zabeel Park gate and then many of them were fixed in other parks, beaches and other locations over the course of the year.

The initiative has been made possible through cooperation between Dubai Municipality in line with the UAE Cabinet's decision to make 2015 the Year of Innovation. Each Smart Palm runs entirely on solar power and offers high-speed Wi-Fi up to a range of 53 metres, supporting up to 50 users at a time. It also has several phone and tablet charging points offering two-and-a-half times faster charging speeds than a regular plug.

What's more, two screens on each palm offer weather information, local news, a navigation application, general city related information—and even a selfie camera [14].

### 6.4.2 Digital Photo Stands

UAE has introduced digital photo stands in few of its public spaces and beaches. The visitors or the beach goers can stand infront of the stand and get their pictures clicked with a background of the location. The pictures are then sent to the visitors via emails.

### 6.4.3 Smart Arrangements for Bikers

Shower and changing rooms at destinations for the bikers in the UAE is encouraging car free travels and encouraging the bikers to bike to the nearest public transport hub/metro station while going to their workplace. To combat with the extreme weather and its impact on the human body, the destinations are getting equipped with shower cubicles, rest rooms, changing rooms and vending machines with water, juices etc.

### 6.4.4 Smart Entertainment System—Open Air Cinema

As the weather warms up and the evenings get balmy, UAE brings the concept of open air theatres and cinemas to the city's top spots to enjoy films and theatres. The outdoor cinema season is in bloom. It's not just catching a good flick in the park but also bringing the entire community together with entertainment, too, so step outside and get watching.

The largest outdoor cinema screen in the UAE, the 20 m x 10 m high-definition screen shows classic films and sport and offers a unique way to enjoy films on a warm evening. The rows of white sofas attract those who are happy to relax with shisha and attractive food in the menue. Movie fans are given headsets to hear the audio—since the venue is essentially a lounge first and a cinema experience second, so music still plays on the speakers—and films are often played with Arabic subtitles. The seating area is comfortable and if the outdoor temperature dips, staff will happily provide you with a blanket. Though this isn't a venue for diehard movie fans—you may prefer a more immersive experience—it's certainly a fun place to chill out and enjoy an al fresco flick.

### 6.4.5 Interactive Boards for Directions

Digital Boards are among the most innovative interactive tools in the recent years and have been working wonders at parks, beaches, restaurants, cafes, lounges etc. An appealing display of locations, menu on a digital signage menu board makes it easily accessible to the customers. A customer can have a look at the entire board, shops, restaurants, themed shops etc. in just a glance and be easily convinced to make a purchase seeing that attractive image of his destination.

## 7 Maintenance of Parks and Green Open Spaces

As per the practice in UAE parks are open spaces are created in limited numbers but larger in size. Ail the parks in UAE are planned to include multiple activities and facilities with ample landscape features. The parks are also facilitated with ample parking and safety security arrangements that can support large scale events arranged within the parks and open spaces.

Limited large open spaces and parks are easy to maintain with the installation of smart gadgets and few control/monitoring stations.

On the other hand multiple of small parks will require more people to maintain and multiple infrastructure facilities scattered at multiple locations.

The developer of the community maintains the parks, open spaces and public realm. The maintenance charges are paid by the community residents in terms of service charges, hence there is an involvement of the government, developers and residents in the creation and conservation of parks and open spaces of UAE.

## 8 Monitoring of the Parks and Open Spaces of UAE

Performance monitoring of the parks and open spaces commences continuously through various ways and means with the judgement from residents, visitors and the users of the facilities. The feedback from various sources are considered for the improvisation of the areas and the facilities within the spaces.

The property developer's role is to select the smart technologies for their development in such a way that the development becomes sustainable, energy saving and attractive for the users. Smart selection of the elements and technologies will also add value to the property with a great chance for high return on the invested amount.

Smart gadgets to obtain user's feedback for all the facilities provided in the public areas, UAE has introduced feedback kiosks for the visitors. The feedback request machine is a part of judging the Dubai happiness index.

## 9 Learning From—Smart Land Scape Architecture in the World

Urban landscape plays an important role in the urban ecosystem and efficient landscaping strategies can lead to better urban design, environment and improved quality of life. At present, the urban landscape is mostly approached as a subset within urban design and the extent to which it can influence the urban environment and quality of life is often overlooked. In this chapter, we aim to delve deeper into this relationship and study the level of influence landscape have on urban dwellers.

The internet and communication technologies has revolutionized most of the industries and this led to the development of smart cities that utilize IoTs, IoEs, IoMs, E-governance and real-time monitoring devices to provide its citizens with a better quality of urban life. These technologies can be incorporated with the landscape design strategies to achieve smart landscape that integrates aesthetics and functionality as well as address various issues such as water management, environmental degradation and climate change.

In the following section, some of the smart landscape designs adopted in cities around the globe are discussed with respect to their success in achieving the set goals.

### *9.1 Case Study of Rotterdam*

In the city of Rotterdam, as an initial step towards achieving smart and sustainable landscape design, a landscape character assessment (LCA) was carried out to study the landscape typology. The landscape design of the city was carried out based on the results of the LCA. Various studies have concluded that the LCA has helped in achieving better urban landscape [4].

Even though, the use of smart technologies in the landscape of Rotterdam is minimal, the depth of LCA has helped in identifying and studying the landscape problems in the city. Such an in depth analysis of the landscape based on typology, urban form, environmental sensitivity and functionality of spaces would result in the formulation of sustainable and smart landscapes.

## 9.2 *Integrated Landscape Initiatives in Practice: Assessing Experiences from 191 Landscapes in Africa and Latin America*

The increasing demand for agricultural land use to increase the supply of food and energy, campaigns for the conservation of environmentally sensitive areas and increasing demand for urban spaces and urbanisation coupled with diminishing resources and the limited supply of land created a conflict in African and Latin American cities. Smart landscaping based on the multi-sector, integrated landscape was thought of as a sustainable solution. This approach emphasizes the promotion of synergies and management of trade-offs among economic, social and ecological dimensions of the landscape, collaborative decision-making processes and supportive market and policy context. Even though, the application of smart technologies was limited, the study concludes that the integrated approach towards landscape has helped in achieving increased productivity, environmental conservation, better policy and planning initiatives along with a better quality of life [5].

This integrated approach towards landscaping can be further incorporated with the concept of smart cities by identifying the various problems in the urban areas related to resource management, citizen satisfaction related to basic services, urban design and urban environment. The internet and communication technologies can provide a platform on which the integrated landscape design strategies can be implemented and monitored.

## 9.3 *Smart Landscape for San Diego County*

The smart landscape grant programme initiated by the San Diego Water Authority along with the Department of Water Resources Proposition aimed at to reduce water demand by improving the efficiency of landscape irrigation. Various landscapes across San Diego was provided with the menu type irrigation facilities based on internet and communication technologies. The scope of the program included application of smart controllers and sensors, quality control, incentives and monitoring. The program was able to reduce the water used for landscape irrigation by 13%. Due to the high initial cost involved in the smart irrigation system, the authority is currently working on developing cost effective smart landscaping systems [15].

Water conservation is one of the most important aspects of smart landscaping. The water requirement for urban landscapes has to be reduced by implementing smart landscaping strategies.

### 9.4 Water Smart Landscapes by US-EPA

The outdoor water requirement account to 30% of the water demand in the United States of America. The study not only emphasize on the use of smart technologies to achieve smart landscapes, it suggests various measures such as choosing plants based on their water needs, group planting based on water demand and maintaining soil quality to achieve smart landscape. According to the study, the term smart landscape is not limited to the use of technology, it includes climate and resource sensitive landscape design.

From the various case studies, it can be noted that most of application of ICTs were limited to water resource management. The scope of application of ICTs in areas such as energy conservation and generation, urban agriculture and developing multi-functional urban landscapes to facilitate smart living through smart landscapes has to be studied further.

## 10 Smart Landscape Architecture-An Integrated Part of Smart UAE

Over a short period of time, large areas in the UAE got transformed from deserts to green lands which include tree planting, creation of public parks gardens and recreational facilities in the cities and along the median and road sides.

The UAE has also converted the natural desert into productive agricultural land with the idea of greening the desert. In UAE more than 330,000 ha forest had been planted to promote soil conservation till 2009 and the agricultural land of UAE has reached 75, 283 ha (as per Abu Dhabi Statistics Center, 2015). This artificial construction of a green nature has been the cause of many environmental issues also.

The UAE has the highest consumption of water per capita in the world, with the domestic consumption of only 13 per cent, while 80 per cent of water is utilized for different greening projects. Sweet water is obtained from desalination of sea water with a huge cost.

UAE is trying to use smart techniques to reduce the consumption of sweet water and also reduce the cost of desalination and yet meet all the necessary water demands for the country. After meeting the demands of the population of UAE, landscaping became the second major priority for maintenance. Some of such smart techniques adopted by UAE is described below.

# 11 Smart Innovations –In Landscape Architecture

## 11.1 Intelligent Irrigation Systems

Smart irrigation strategies and water efficient irrigation networks, such as sensor controlled and drip irrigation systems, reduce water consumption by green spaces. Zone irrigation, the use of mulch and soil amendment techniques, can also help increasing water retention capacity of the soil and reduce evapotranspiration.

Another simple method to reduce water consumption is to avoid watering green spaces in the middle of the day and scheduling lawn irrigation for early morning or evening hours. Using drip irrigation systems or a cycle-soak method, instead of sprinklers, is also an attractive option for achieving greater water efficiency and improved water penetration.

Middle east area is plagued by high irrigation water losses in sandy soils as the plants have little time to use the irrigation water before it percolates to the water table.

Such interesting innovation is the development of water retention additives that can potentially cut the irrigation water by as much as 50% [15–17].

## 11.2 Distribution of Green Space to Various Levels

### 11.2.1 Roof Top Garden in Dubai

Dubai has come up with successful gardening effort above the ground and on various vertical levels. The roof top and podium level gardens are becoming part of large scale hospitality developments. The water consumption is controlled by using special soil in place of sand and watering is done through smart irrigation systems.

Some of the hotels are growing fruits and vegetables that are being consumed by the hotel users.

Oak Hotel & Suites in Tecom, Dubai have cultivated such food forests, which that act as show garden for the propagation of fruit, vegetables and herbs grown in adherence with international standards for organic produce.

The garden has brought together individuals from many backgrounds, professions and countries, and as it has developed, it has also attracted birds and bees to the site. A small garden party of thanks was held at the weekend, where locally sourced produce was served and volunteers and other guests were able to stroll around the garden, to view the 50-plus types of plants and vegetables being grown there.

### 11.2.2 Vertical Gardens in Dubai

Vertical gardens, or living walls, are springing up across the UAE, as innovative new technologies, coupled with increased horticultural understanding, have expanded the potential design permutations of this living art form.

Commercial office and showroom interiors feature living wall by-Vertical Garden Design. The feature includes rhipsalis, along with more-compact begonias, peperomias and streptocarpus.

A wellness center in Jumeirah features a moss vertical wall by Verdevip. The wall consists of moss harvested from bio–sustainable areas, and is as low maintenance as it gets, providing that installation is made in environments where there's a humidity level of 40 per cent and no direct sunlight. This makes it ideal for gyms, spas, pool areas and darker rooms. The tiny plants are treated with glycerol and food colourant before being attached to panels.

There are gardens consisting of 1,900 cascading plants, including green ferns, silvery aglaonema and the violet hues of-anthurium.

There are 24 vertical garden installations at the Sofitel Dubai The Palm Resort and Spa, which in total cover an area of 600 square metres where koru, a fern symbolising the embodiment of new life, is one of 120 species deployed at the resort.

## *11.3 Garden on Mat*

A thin, natural fibre mat created to encourage the bloom of rooftop gardens by radically reducing cooling bills due to the double insulation they provide for homes and high-rises.

Green roofs are recognised worldwide for helping to reduce urban heat island effect, the term given to the overheating of urban and suburban areas due to increased paved, built-over and hard surface areas.

A Green Mat is made of coconut and date palm parts, that traps a lot of water in small amounts of soil. The flat brown mesh is spread where the natural vegetation can grow and is a way to grow with limited water consumption. The lightweight Green Mat looks and feels like a brown, wiry doormat. Shallow five centimetre layers of soil sandwiched between swathes of Green Mat help trap water and keep soil moist, reducing the need to irrigate. It can reduce the weight of a traditional green roof by 500 kilograms. Priced at Dh25 per square metre, return on the investment in terms of water saved is six months.

In 2008 Dubai Municipality launched the 'Green Roofs Initiative' and announced that at least a third of the surface area of roofs should be green. The new green building code will make it mandatory for all new buildings to have a roof garden.

Green roofs provide double insulation for the buildings, save on air conditioning, reduce the carbon footprint and can save up to 60 per cent of water compared to a regular garden.

### 11.4 Use of Zeoplant

is a natural, moisture–retaining soil additive, and halves the necessary quantity of irrigation water. Added to the soil in a one-time application, it improves the soil structure by helping it to hold and retain irrigation water, making it available for plant roots. So effective is the product in the Middle East's climatic conditions that Zeoplant has been used in sustainable landscaping and agricultural projects in the UAE, Bahrain, Qatar, Saudi Arabia, Kuwait and Oman [18].

### 11.5 "Living" or "Green" Walls

These low-maintenance, vertical gardens covered with vegetation fulfil the primary aims of any landscape gardener: aesthetic appeal combined with environmental friendliness. Using high-tech irrigation systems, these green facades can be grown in almost any urban habitat, and are gaining a reputation as an effective feature of sustainable landscaping. Aside from their striking beauty, living walls bring a variety of benefits, from improving energy efficiency by enhancing insulation and reducing carbon emissions to acting as noise and pollution barriers, as well as climate–control agents.

The concept of green walls is gradually taking off in Dubai, with the arts-and-culture centre The Pavilion in Downtown Dubai boasting a 10-metre-long vertical interior garden, featuring about 2,500 plants. Elsewhere in the city, the Novotel hotel in Al Barsha has a six-storey green wall containing approximately 27,000 plants.

## 12 Imposing Rules for Smart Use of Water Resource

### 12.1 Steps Taken by the Municipalities

Dubai Municipality (DM) has capped the use of treated water supply to 5 litres per sqm, per day. Dubai—New rules on landscape watering are in the pipeline to ensure garden lovers in Dubai don't waste irrigation water while going green.

The Dubai Municipality (DM) is planning to put a cap on the use of treated wastewater supply from DM for irrigating private landscape projects and residential gardens.

The rule is expected to prevent people from leaving their sprinklers and hoses on for longer than required time and be cautious about water usage.

In a related development, Emiratis using treated water supply from the municipality to water their gardens and landscape in their villas will be charged a fixed monthly fee from this month.

The old method of metering will be replaced by a flat rate as per the plot size that is applicable for treated irrigation water supply, whether the water is used or not.

## 12.2 Big Savings Through Usage of Smart Renewable Energy Sources

On its part, the municipality has implemented sustainable initiatives in Dubai parks that have helped the civic body cut down the carbon footprint and save millions of dirhams.

The civic body so far saved around Dhs 3 million in power bills by reducing the power consumption by 5.5%.

Since 2012, a lot of initiatives are taken to make the parks and landscape projects sustainable.

The steps taken by the municipality include using solar energy for lighting and watering, planting more local trees, using recycled materials for hard landscape, using tiles, play equipment and other materials as per green building standards.

## 12.3 Drones for Survey of Landscape Areas

Reducing water wastage in landscaped areas and checking the upkeep of plants are turning out to be the jobs done by drones.

Taking innovation to the field of landscaping, drones are being deployed to survey and monitor large-scale landscape projects.

Number of plants are there in the project, adequacy of water and fertilizers, monitoring plant decay and such details can now be captured by drones with much less effort and time.

Drones can also be deployed to avoid water wastage in large-scale landscape projects.

It can monitor the temperature, water supply or whether the supply is more than what is required. The action can include areas to be treated with soil additives without digging up and checking samples.

The precise images taken by drones with a colour coding system help professionals get a good interpretation of the surroundings.

However proper research is required in the field of environmental cartography. Aerial cartography is a powerful instrument with that helps in monitoring the quality of all the trees, one by one, from above.

## 13 Smart, Safety and Security Devices for City and Its Public Realm

Safety and security is another important aspect for maintaining the precious land scaped areas and areas with large scale events within the cities [19, 20].

There are technologies to maintain security and boost productivity in police work. Some of them are:

### 13.1 *The 'Falcon Eye' System*

Used by Abu Dhabi police is a complete central system, which called the falcon eye system that receives a live feed from visual surveillance equipment installed on site and displays data in an integral system that provides warnings and enables quick access to events and incidents [21].

The system controls roads by monitoring traffic violations such as the use of hard shoulders, illegal parking and the misuse of roads and dealing with traffic accidents live.

Useful links:

Dubai's flying taxi to soar by year-end—Khaleej Times

Driverless Mini Police Cars to Patrol Dubai—Gulf News.

### 13.2 *Flying Bikes*

In 2017, Dubai Police unveiled a model of a flying motorbike, Hoversurf that can enable one police officer who pilots the bike. However, it can also fly through remote operation. It is being considered for use during emergencies.

Hoversurf works on electricity and can fly at five metres height. This feature is useful during emergencies and heavy traffic. The flying bike can travel up to a distance of 6 kilometres. It can fly for 25 min and can carry up to 300 kg of weight at a speed of 70 kmph.

Useful links:

Dubai Police unveil flying bike, robotic vehicles—Gulf News.

### 13.3 *Engaging Public in Maintaining Security*

The UAE Government recognises that nation's security is a collective responsibility and that individuals can potentially help in realising security and preventing crimes.

Therefore, it introduced some security services that enable people to communicate and report any suspicious or harmful incidents directly to the authorities.

### *13.4 Aman Electronic System*

Aman electronic system in Abu Dhabi enables individuals to raise security concerns they encounter. The information is used to assist the security agencies in preventing specifically planned crimes and locating criminals [22].

### *13.5 Al Ameen Service from Dubai Police*

The Al Ameen service enables people to contact the police with complete confidentiality, round the clock, about any problems or concerns they may have. The service studies each of the cases brought in. Some of the issues that were reported are [23]:

the presence of blacksmith shops that sell knives to teenagers

the receipt of fraudulent text messages by mobile phone users.

### *13.6 Al Najeed Service from Sharjah Police*

Sharjah Police introduced the Najeed service. Through this service, you can report information and tip-offs about crimes or any other threats posing danger to the stability of society.

### *13.7 My Security Is My Neighbour's Security*

My security is my neighbour's security is a programme for the residents of villas and houses in Sharjah under which residents are encouraged to report any suspicious movement or activity in their neighbourhood and prevent crime [14].

## 14 Policies

Implementation of the policies related to sustainable and smart development into the old and new development was a challenge. UAE is a mix of traditional and contemporary development and introducing the guidelines for both the development typologies was a greater challenge [24].

The region has proved to be smart in the implementation of guidelines are made available and accessible through the municipality and authority websites and the incorporation of the mandatory regulations are periodically followed up by the Authority during the property licence renewal. The approval process of the new property development is complete after a thorough review of the considerations of the regulations.

## *14.1 Green Agenda*

In January 2015, the UAE Cabinet approved the UAE Green Agenda 2015–2030 as an overarching framework of Green Economy actions. The Green Agenda consists of 5 Strategic Objectives (Competitive Knowledge Economy, Social Development & Quality of life, Sustainable Environment & Valued Natural Resources, Clean Energy & Climate Action and Green Life & Sustainable Use of Resources). It has 12 main programs and 31 sub-programs. The Emirates Green Development Council, formed in June 2015, coordinates and monitors the implementation of the Green Agenda [25].

## *14.2 Green Leadership*

UAE continues to power ahead with its innovation strategy. This time it's pushing for a global leadership role in the green economy with openly stated and highly accountable aims [25].

In December 2017, the emirate hosted the Green World Awards, which celebrated public and private sector environmental best practice. The right to host the prestigious event, a product of the global NGO—The Green Organisation, came after the Roads and Transport Authority (RTA) claimed the 'Champion of Champions' title for its energy practices in 2016.

## *14.3 Creation of Green Building Guidelines*

Dubai Municipality launched the new Green Building Regulations & Specifications for Buildings in the Emirate of Dubai. These form a baseline and are not a rating system such as the USGBC or Estidama Pearl Rating Systems, conditions to be followed in the green building standards and specifications, and Structure in the Emirate of Dubai [26].

Dubai 2020 Urban master plan aims to build an advanced global city and a vibrant regional gateway to the world. The plan which has been developed by government is in cooperation with various governmental authorities and major developers, adopts a flexible, sustainable and competitive approach to guide the urban development to 2020 and beyond. The key objectives include the efficient utilization of existing infrastructure, increasing investments opportunities, rationalizing a well integrated land use planning and mobility network, protecting key economic assets, preserving the environment, diversifying energy sources, being responsive to the socio-economic transformation and improving the legislative and governance framework to streamline planning process.

## 14.4 Promoting Smart Innovations

As part of its vision 2020, the UAE government has earmarked billions to develop urban greenery, through landscaping and horticultural projects, in a bid to transform the emirate into a "green paradise".

For a modern 'smart city' sustainable ecological landscaping is an important theme. "The use of 'smart' solutions, including-innovation within the landscape industry, and the increased use of technology, are making it easier for urban planners and environmental specialists to-implement sustainable, ecological landscaping initiatives.

UAE government has also announced plans to expand the city's irrigation network, which will in turn support its landscaping projects, as will the use of materials such as special sand and stones to reduce water consumption and prevent groundwater from filtering up and spoiling plants.

## 14.5 Promoting Global Level Meet in UAE

As UAE continues to witness a rise in construction projects, there is also a marked growth in the investment of green open spaces, golf courses, sports stadiums and public parks & gardens. A part of its efforts to support and achieve the sustainable growth of green spaces and encourage a better quality of life the government is investing in promoting technologies for sustainable development and also increasing the interaction between the experts of this field to share their knowledge and decipher the same to the professionals.

A yearly event called the Middle East Smart Landscape Summit takes place in Dubai and Abu Dhabi that involves the professionals from government, private and academics to come together and share their work, innovations, solutions to various landscape development and studies. As the region's leading event on landscape solutions, the summit aims to bring to life the trends and developments in the landscape industry.

## 15 Funding for Smart Development

There are organisations headed by H.H. Sheikh Hamdan bin Mohamed bin Rashid Al Maktoum, Crown Prince of Dubai and Chairman of the Executive Council that aims to place UAE amongst global cities leading the transition to the global green economy.

The municipalities are fully committed to supporting and realising this goal.

Dubai GEP is the first multi-stakeholder and cross-sector partnership to sustain focus on a green and low-carbon economy in the Middle East, with the founding members coming from both public and private sectors. Early in 2015, the Dubai Supreme Council of Energy (DSCE) formalised a strategic partnership with DM to promote Dubai's vision, strategy and achievements to an influential global audience and forge world-class partnerships to help build capacity through projects, event platforms and green trade missions, amongst other activities and programmes, in line with the Green Growth Strategy.

## 16 Monitoring Performance

UAE performs a continuous assessment of the facilities provided for the people of UAE including citizens, residents and tourists visiting the country. A unique method called happiness index method is introduced at every operational centers such as public facilities, government facilities where people can express their satisfaction level or experience through the installed meter with options or drop boxes.

The assessment of the performance of the facility and improvisation is decided based on the experience/observation of the users fulfilling the vision of HH Sheikh Mohammed Bin Rashid Al Maktoum to make the city-THE HAPPIEST CITY ON EARTH [27].

## 17 Future Strategies

Environment Vision 2030 was developed in close coordination with the General Secretariat of the Executive Council and Department of Economic Development/Abu Dhabi Council for Economic Development to ensure integration among the environmental, economic and social visions, the three pillars of sustainability [28].

Vision 2030: At the request of the Abu Dhabi Executive Council, the Environment Agency—Abu Dhabi (EAD) is developing the Abu Dhabi Environment Vision 2030 policy agenda as an integral part of Abu Dhabi Government's holistic policy agenda framework encompassing the driving Economic, Environmental and Social Agendas that together will steer the Emirate towards sustainable development. Environment

Vision 2030 has been developed through an extensive consultative process involving stakeholders from across the board with the intent that it serves not only to protect the environment but also to complement the economic and social agenda for Abu Dhabi, as well as build upon existing environmental initiatives. A Vision Statement has been drafted that summarises the objectives and proposed policy programme of Environment Vision 2030: 'To preserve and enhance Abu Dhabi's natural heritage, while acting as a regional leader in the efficient use of resources and contributing to a better quality of life for all' [20, 27].

The drive for more sustainable built environment in the United Arab Emirates (UAE) received a significant boost in the period 2006-2008 supported by new legislations from the government and increased interest from major stakeholder especially municipalities. The absence of locally developed frameworks for evaluating sustainable design of buildings has created opportunities for the introduction of foreign ones.

LEED has become popular in industry and has even influenced the current attempts to develop a local framework. The paper will report the findings of a research that, first, evaluated the effectiveness of LEED framework in dealing with local issues. Secondly the paper will examine two case studies where LEED was used in only one of them while both cases aimed to ensure the project development is sustainable. The analysis shows that "responsible design" produced more effective solutions for sustainability.

UAE will continue to expand and conserve the green open spaces though various innovative efforts and will set an example for the other countries.

## 18 Conclusion

The purpose of innovative technology is to simplify life and Dubai is moving toward the goal of creating "an intrigued" example of the whole smart concept.

In moving towards the change, the building efficiency and sustainability play a vital part of the entire smart city concept—but there are also individual national, economic and cultural needs which have to be met by technological developments. In many larger projects so much energy is being wasted and that is something to avoid. Hence there are lessons to be heeded.

Adaptability and malleability are other key factors in smart design and the future projects need to strike a balance and be flexible between the built and open environment creation [29].

Smart technology should also enhance the quality of people's lives and reduce the consumption of resources. Dubai Municipality is going about these aims the right way with ideas such as the increased use of solar panels to generate energy.

A smart city starts with mobility and creates a place where individual feels safe.

Data storage, data protection and data privacy is also a part of smart design.

Hence a holistic approach with a collaborative movement involving all stakeholders– including the research and development community, government authorities, developers and the general public is necessary in forming and running a smart city.

Undoubtedly the benefits of a society that is seamlessly synchronized with technology will not only enhance but also evolve our living environments.

Ultimately, the factors that enabled the government to expedite its implementation of development are summarized as follows:

- Adopting a clear vision based on common universal value
- Applying methodological approaches in assessment of weaknesses,
- Analysing gaps and designing roadmaps
- Openness to learning from international practices
- Solid partnerships with local and international private sector entities
- Seeking international standardization in measures of evaluation
- Acting entrepreneurially beyond hierarchies
- Infusing collaboration across government through soft and hard measures
- Piloting, showcasing ideas and leading by example
- Agility in restructuring government functions and bodies as needed.

## References


1. Paul S (2010) Landscape Planning—preservation, conservation and sustainable development. Department of Landscape, University of Sheffield
2. Strong signals: growing use of public Wi-Fi hotspots. Available at www.acma.gov.au/researchacma, 26 Nov 2014
3. Lifeline for modern cities, by Tatiana Antonelli, Founder and Managing Director of online environmental directory Goumbook, August 2012
4. Fairclough G, Herlin IS, Swanwick C (Eds.) (2018) Routledge handbook of landscape character assessment: Current approaches to characterisation and assessment
5. Hart AK, Milder JC, Estrada-Carmona N (2014) Integrated landscape initiatives in practice: Assessing experiences from 191 landscapes in Africa and Latin America
6. Landscaping with native plants in the UAE: a review, by Hasnain Alam, Jabar Zaman Khan Khattak, Shaijal Babu Thru Ppoyil, Shyam Kurup, Taoufik Saleh Ksiksi, year 2017
7. Zilliacus A (2017) Low-cost techniques to activate underused urban space. 23 Feb 2017
8. Solar power palm trees with wifi and charging points, UAE. Available at https://www.designboom.com/technology/solar-smart-palm-trees-wifi-dubai-08-01-2017/
9. Landscape and roadmap of future—internet and smart cities. Available at www.fireball4smartcities.eu
10. Environmental guideline for smart cities, Ministry of Environment, Sustainable Development, and Disaster and Beach Management, July 2015
11. Smart Dubai—happy living. Available at https://smartdubai.ae/en/Pages/default.aspx
12. Community facility standards, Dubai municipality. Available at https://www.dm.gov.ae
13. And Edward T, McMahon JD. Available at http://www.sprawlwatch.org
14. Pfledderer S Smart moves: smart home technology. 7 May 2015
15. Greer KA (2005) Habitat conservation planning in San Diego County, California: Lessons learned after five years of implementation
16. Bennett R, Hazinski H (1993) Water-efficient landscape guidelines

17. Smart landscapes for California (2005) California Urban Water Conservation Council
18. Zeoplant I save water - save environment - save cost, available at https://www.zeoplant.com/
19. Landscape and road map of future-Internet and smart cities(M24), April 2012
20. The landscape approach, The concept, its potential and policy options for integrated sustainable landscape management, 9 Feb 2015
21. Levine WS (2018) Falcon eye system- control system applications
22. Aman electronic system, available at http://www.aman-me.com/
23. Al Ameen Service, available at https://alameen.gov.ae/
24. Environmental Guideline for Smart Cities (2015) Ministry of Environment, Sustainable Development and Disaster and Beach Management, Environmental guideline for smart cities
25. Benedict MA, Edward T, McMahon JD Green infrastructure: smart conservation for the 21st Century, The conservation fund www.sprawlwatch.org/greeninfrastructure.pdf
26. Green building guidelines, UAE
27. Sustainable landscape. Solar Landscape, Environmental Design Collaborative. Available at edc@sentex.net
28. Sustainable landscapes. edc@sentex.net
29. Jung JG (2015) The art of landscape in smart cities. 1 Dec 2015
30. United Arab Emirates green economy report (2016) 2nd end.
31. Smart landscape practice in UAE. Available at https://gulfnews.com/smart-summit-for-landscaping-1.1328583
32. Maranhao D (2015) Water-smart landscapes start with WaterSense
33. Green economy report (2016) 2nd edn. United Arab Emirates Ministry of Climate change and environment
34. Ariana Z Low-cost techniques to activate underused urban space
35. Abu Dhabi public realm design guidelines. Available at https://www.dpm.gov.abudhabi/

# Conclusion

# International Collaborative Research: "Smart Living for Smart Cities" and Conclusions of Cities Case Studies

**T. M. Vinod Kumar**

**Abstract** This chapter has two parts. In the first part, the organizational details of the international research collaborative project "Smart Living for Smart Cities" is discussed. In the second part are presented in consultation with the team leaders of the city study, their general conclusions of the study.

**Keywords** Organization study · Results

## 1 Smart Living for Smart Cities

This "Smart Living for Smart Cities", Book is an outcome of international collaborative research aimed at developing the Design and Practice of Smart Living for Smart Cities. The Smart Living calls for an active role of Local Community transformed into smart community, Local higher educational Institutions, Chamber of commerce and industries, and Government and hence its management is a joint and synchronous effort of E-Democracy, E-Governance, ICT and IOT system in a 24 h 7-day framework in all aspect of life in any smart cities. The ecological system under transformation and cultural background of inhabitants with a long past have a greater bearing on the smart living which has been detailed out in Chapter "Smart Living for Smart Cities".

The smart living is a practice that uses information and communication technologies, Internet of Things, Internet of Governance (E-Governance) and the Internet of People (E-Democracy) along with conventional management tools to realise smart living for all. It is a coordinated, effective and efficient information-based management system for the smart community for daily living, human development, and conservation that improves ecological systems, strengthen the cultural system, improve social welfare in an equitable manner without compromising the sustainability of development ecosystems and stakeholders' cultural values. This book will present many city cases studies from some countries centred on one or many specific aspects

T. M. Vinod Kumar (✉)
School of Planning and Architecture, New Delhi, India
e-mail: tmvinod@gmail.com



T. M. Vinod Kumar (ed.), *Smart Living for Smart Cities*,
Advances in 21st Century Human Settlements,
https://doi.org/10.1007/978-981-15-4603-7_6

of living in study city. The Smart Living energises, reorganises and transform the living style existing to a smart living embracing highly responsive ICT and IOT, creating sustainable and harmonious living, responsibly sharing this effort, by smart communities.

These city studies conducted in this book start with a background Chapter "Smart Living for Smart Cities". The state of the art of smart living studies has a history of never-ending progress and refinement to create a view which often faces alarming human issues but generate timely intervention often automatically and timely action towards these issues. Our ancestors have kept living in better shape than us. Now many communicable diseases, mental ailments, and lifestyle diseases emerging today which was absent earlier. Without a system for timely intervention, we would have been lost among many species that is no more living. There was a strong cultural system practised by households and communities among our ancestors based on a way of life of many religions which maintained a robust living in a sustainable manner. This explains our very existence.

Smart Living for Smart Cites can be conceived and developed emphasizing the sum of integration of internet of enabling ICT and IOT technologies for the smart living, embracing a carbon-neutral living style monitored in real time with a transportation system that performs intelligent and smart mobility based on real-time information drawn from big data using electric energy for movement replacing fossil fuel with its pollution and disease. The authors of this book have adopted only one simple and dependable way of looking at Smart Living and smart living governance through smart activation of Smart People. Smart People need not embrace Capitalism since they have found that without much capital, not owning a brick and mortar store, or a mall, they can be part of Amazon or eBay seller which is the biggest market place in the world sharing the ICT-enabled marketing and logistic system at low marginal cost and price. In the same way, they can also be a part of the largest taxi services like Uber with only one taxi at his disposal while Uber does not own even one taxi. They can be part of Air B and B without owning a hotel defeating the central concept of Capitalism. Since Uber just provides location-based computer and smartphone platform for the taxi service it does not have a huge and very expensive (with very limited benefit) bureaucracy that dictates movements, unlike Socialism and Communism. In a similar way, we must find ways and means for smart living by smart community suited for local needs. Since real-time information and big data guide all these economic activities, socialism and its consequent ills of proliferating and expensive and mostly corrupt bureaucracy which can be oppressive and disgusting for taxpayers with no value added to the urban economy are no use for the Smart city economic development and smart living. In simple term, data-based, or well-informed living can be what smart living ultimately based upon.

During the colonial period, everything about life was centred around the bureaucrats of the colonialist with a Stockholm syndrome. Even when many countries became free from colonialism this dependency on bureaucracy was evident as if bureaucrat regulator can give leadership to create a good living which by the very job of the bureaucrat cannot. This is a false image a part of Stockholm syndrome. At best, he can implement policies of the elected government enshrined in legislation in close cooperation with people with a history of more failures than success which is due to lack of innovations. This does not mean any regulation or no law and order in Smart Living. Like tax compliance by the electronic filing of Goods and Service Tax, income and wealth tax, Smart E-Governance is there to replace age-old and dysfunctional bureaucracy enshrined in a brick and mortar building in a prominent place in the city, a legacy of the sixteenth century in many colonial countries. We consider Smart People and smart community and their E-democracy as all-powerful and capable building blocks of Smart Cities for Smart Living replacing capitalism and socialism as well as the largely dysfunctional bureaucracy of socialism at one go replacing with smart and sharing living. For Smart People, there should be opportunities in Smart Cities for continuing education and skilling for Smart People to make them smarter today compared to yesterday in a rapidly urbanising world with new skills. This calls for the closer relationship of the community with academia which does not exist today. Potential Smart People can be everyone in a city irrespective of their wealth, educational qualification and social background and, therefore, an inclusive concept since all of them have a constructive role to play in Smart Cities to create the smart living designed by themselves. They can be below the poverty level or above, which does not matter but all of them should have the wish to be Smart People and can be part of never-ending learning mode to be smart. Smart People through their E-Democracy and E-Governance plan, design and govern the Smart Living system in Smart Cities. Smart People are the creators, governors, regulators, managers and maintainers of Smart City Living. This replaces the expensive bureaucrats. The required Smart Cities technologies which are ICT-enabled can easily be designed by Smart People's creativity in collaboration with academia and business community, and prototypes are made in Fabrication Laboratory (Fab Lab) is in Smart Cities for later adoption and mass production and use. Being the creators of the Smart city technologies, Smart People can maintain, repair, innovate and evolve the existing technologies to more cost-effective and functionally superior, next-generation technologies which can be shared profitably with other smart cities for creating a smart living the international goal. No one is running away from Smart People's creation and redevelopment of next-generation Smart city technologies like some international firms.

## 2 The International Collaborative Research Projects on Smart Cities

This book is the seventh and eighth in two volumes, in a series Professor T.M. Vinod Kumar conceived, coordinated, implemented and edited about articulating the various roles of Smart People in Smart Cities.

The first book entitled "Geographic Information System for Smart Cities" [1] was aimed at creating a comprehensive self-awareness of city functioning every second and every day in real time which is the foundation of Smart city. Geospatial technologies, sensors and analytics can be used to reach the awareness and use it in real time for various types of use by Smart People. How it can be used for a variety of urban issues commonly observed globally is what that book is all about.

These Smart People thereby progress towards their self-directed goals, such as they demand Smart Living, as well as Smart Living and Smart Economic Development. They aspire to the highest level of quality of life in Smart metropolitan or megacity city living which they can very well afford and can expand many folds the economic development opportunities to satisfy higher income and employment needs to sustain Smart People. No smart person in a city is an island or elite, but they share a common destiny and common urban space, urban realm, and social and physical infrastructure. Government as the regulator is required that none of the Smart People is denied of all city provides for irrespective of their income level and social status or they are above or below the poverty level. Hierarchy of government exists in a city, but their governance needs to be for a Smart city that is fully aware of itself every second and as against Government who comes to know about the issue when a case is filed in the court which takes many decades to get a final judgement. The existing governance systems are obsolete being a product of sixteenth-century or earlier designed for colonial rulers, built on the model of East India Company's administration in India or elsewhere which cannot be used or Smart city economic development and for the smart living creation. However, those who aspire to live in Smart Cities are in the twenty-first century and no more part of an exploitative empire under the iron hand of a colonial administrator. Therefore, the twenty-first century Smart Cities require Smart city e-governance system that was the subject matter of the second book entitled "E-Governance for Smart Cities" [2]. This book is all about E-Governance in Smart Cities in action. It is divided into three parts, State of the Art Surveys, Domain Studies and Tools and Issue of E-Governance in Smart Cities.

The third book in this series is, "Smart Economy in Smart Cities" [3]. This book explores possibilities for rapid change in the income level and employment opportunities of those Smart People below or above the poverty level in a Smart city, and to make the NDP growth rate to a desired higher level consistently for the next many decades. Then, the current trend of urban local economic development is required to be converted to Smart city Economic Development. A, 10% NDP growth rate envisaged for the next three decades in India and many other countries can only be realized through Smart city Economic Development. The projection that Indian GDP in PPP will be the second largest in the world after China in 2050 can only be realised through smart metropolitan or megacity economy. Smart Cities and the related conceptualization boast of the Smart Economy but not much has been systematically researched or documented about it so far. This calls for a study of many cities across the world to document what constitutes a Smart Economy. There are two groups of cities being studied in this book. Some of them have been designated as Smart Cities by learned societies, but others are not but aspire to be Smart Cities. These call for different approaches to research design and studies. It was seen from case studies both these cases in different countries emphasize different approaches, establishing that there are no cookbook solutions. The cities being studied in this book are spread across several major continents and regions, including

North America, Europe, Africa, Indian subcontinent and East Asia. They are Ottawa in Canada; Stuttgart in Germany; Bologna in Italy; Dakar in Senegal; Lagos in Nigeria; Nairobi in Kenya; Cape Town in South Africa; New Delhi, Varanasi, Vijayawada, and Kozhikode in India; Hong Kong in China, Cape Town, Dakar, Nairobi and Lagos in Africa.

The fourth book in this series is "E Democracy for Smart Cities" [4]. The world over, participatory democracy is worshipped and preached but what is practised is representative democracy at the city level and beyond. It is believed that in meta cities, megacities and metropolitan cities, an only representative democracy with elected representatives will work. However, democracy practised in small cities like Athens in Greece, and the Licchavi Republic in India in ancient times and many parts of the world documents face-to-face democracy in practice. In these cities, everyone in a city sat together and jointly decided on all aspects of the city during war or peacetime. Citizens not only participated in decision making but acted together as one government and even as an administrator for a task and as a regulator. There were no permanent administrators then. With the advent of ICTs in Smart Cities of the twenty-first century, it is possible to go back to the face-to-face democracy in meta city with 20 million-plus population, a megacity with 10 million plus population and metropolises with one million plus population that, by any measure, is much superior to representative democracy. We do not want middlemen the fourth book is all about E-Democracy in Smart Cities in action. It is divided into three parts, State of the Art Surveys, Domain Studies and Tools and Issue of E-Democracy in Smart Cities.

The fifth book in the series is "Smart Metropolitan Regional Development: Economic and Spatial Design Strategies" [5]. Here these cities however large these cities may be, need to be converted to smart metropolis using the specific design of economic and spatial strategies and not by purchasing smart technologies alone. The city studies for the "Smart Metropolitan Regional Development" result in many insights on many smart spatial and economic strategies using the Internet of Things, Internet of Democracy and Internet of Governance oriented to the specific issue of a town and its potential; taking into consideration that the Smart metropolitan city is an integrated six systems in which Smart Economy is an integral part. It can relate to Smart Mobility, Smart Living or Smart Environment. Based on the elaboration of Smart metropolitan city System, if one must develop any metropolitan-, region, then a country-and city specific economic and spatial design strategies for a Smart metropolitan city, must be designed based on a local ecological and cultural system of the city and not a type universal design. Location-specific and culturally acceptable economic and spatial strategies can be locally evolved, governed and managed. This is the only way local culture will find expression in the Smart metropolitan city using specific economic and spatial strategies by utilizing local, creative talents of smart people in many institutions in Smart Cities. There are 16 cities being studied in this project namely Pittsburgh in USA, Stuttgart in Germany and Naples in Italy, Dakar in Senegal, Conakry in Guinea, Abuja in Nigeria, Johannesburg in South Africa and Nairobi in Kenya, Ahmedabad-Gandhi Nagar, Bangalore, Chandigarh, Jaipur, Kozhikode, New Delhi, Surat in India and Hong Kong and greater Pearl River Delta Region from China.

The sixth book in this series is this book that was published in 2019 by Springer-Nature is entitled "Smart Environment for Smart Cities" as a product of international collaborative research. This book is aimed at developing the Design and Protocol and Practice of Smart Environmental Resources Management for Smart Cities. Environmental Resources are common proprieties where an active role of Government and People are required and hence its management is a joint and synchronous effort of E-Democracy, E-Governance and IOT system in a 24 h 7-day framework on any resource in any smart cities.

The smart environmental resources management is a practice that uses information and communication technologies, Internet of Things, Internet of Governance (E-Governance) and Internet of People (E-Democracy) along with conventional resource management tools to realise the coordinated, effective and efficient management, development, and conservation that improves ecological and economic welfare in an equitable manner without compromising the sustainability of development ecosystems and stakeholders. This book present many city case studies (Hong Kong in China, Ahmedabad, Gandhi Nagar, Chandigarh, Kozhikode, New Delhi, Patna, Surat, Gurgaon from India, Yokohama in Japan, Nairobi in Kenya, Ibadan in Nigeria, and Dubai in UAE), that is centred on one or all environmental resource each in a city.

## 3 The Seventh and Eighth Book on Smart Living for Smart Cities

The editor and coordinator of the book series T.M. Vinod Kumar and many authors who participated in the earlier five books felt that there is a gap in knowledge about Smart Living in a smart city. Funding for such collaborative research project like all earlier book was another issue. Universities and research centres dominated in collaborating these six-smart metropolitan city research projects. We also found that along with Universities, some not-for-profit national and international networks and institutions, city governments and regional governments in certain countries also came forward to participate in this collaborative research programme. The editor and coordinator of the project again felt that this international project shall not seek any external funding other than the internal resource mobilization from within the participating universities (Fig. 1).

This book is expected to present many city case studies (Hong Kong in China, Ahmedabad, Surat, Mumbai, Kozhikode, Thiruvananthapuram, Vijayawada, Leh, New Delhi, Chandigarh, and Kozhikode, from India, Lagos in Nigeria, Galilee Region in Israel, and Dubai in UAE, St. Louis and Pittsburgh in USA) that is centred on different aspects of living in each city.

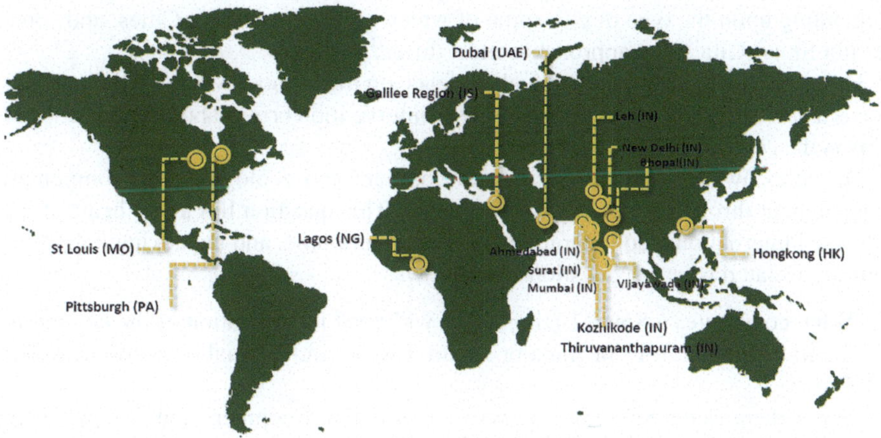


**Fig. 1** The location map of the studies conducted in this book. *Source* Bulletin 1

**Fig. 2** Typologies of research institutions participating in this book

## 4 Design of the Collaborative Research Programme

Research Collaborations worked out is purely voluntary and without any financial support that binds a project together. Since collaborators are universities, Government, research institutions, professional networks and not-for-profit associations from six countries, complete independence for pursuing the research was there, free of the baggage of ideologies of granting organisation. They need not accept existing smart cities policies of study cities in their research. Coordinator Editor of the project has no financial or administrative control over any institution participating in the project since he was not in receipt of any grants and did not distribute it. Typologies of the institutions involved in this international project are given in Fig. 2. All these autonomous institutions are guided by the highest standard of scholarship and timely completion of research and publication.

## 5 Research Questions on Smart Metropolitan Regional Development

The kind of collaboration in this international research project requires that all participating institutions shall formulate their own research questions and research the methodology which is of use to the country where these study cities are located.

Depending upon the type of city some of which are leading Smart Cities, and some are not Smart Cities, the approaches must differ?

However, the paucity of empirical evidence on the Smart Living for Smart Cities opens a new area of research: What strategy intervention brings about Smart Living? This is the central focus of the book.

Do cities awaken social, cultural development and ecological (environmental) management through smart city development? This question lies at the heart of the proposed international collaborative research programme, and unpacking it gives us four interrelated research questions, as follows:

1. What constitutes a Smart Living? This will need identification of the key ingredients and their role in making Smart Living and Spatial changes in Smart Cities.
2. What changes the Smart Living brings to social development, cultural preservation, heritage conservation and ecological management? This calls for understanding the inter-linkages between Smart Living in Smart Cities on one hand, with social development, cultural preservation, heritage conservation and ecological management on the other.
3. How and what processes facilitate the changes to smart living? Do cities bring to social development, cultural preservation, heritage conservation and ecological management? These may include:

   (a) innovation–diffusion (by ICTs and other modes),
   (b) spatial planning,
   (c) sectoral planning (including economic, social development, cultural preservation and ecological management),
   (d) heritage conservation and management plan and
   (e) institutional and governance processes, among others.

4. How and what changes can be brought to improve the processes to achieve improved/optimal results? These changes related to the various processes as mentioned in Research Question 3.

A deeper understanding of changes in the social, cultural and ecological system of the Smart city with the advent of Smart Living and Smart People for smart city development is the focus of study. This research programme and the institutions selected for this purpose as academic collaborators are an effort to address this research gap.

## 6 Scope of Research

The following outlines the areas that may be covered when conducting research under the "Smart Living" programme. This is an indication only, and it is left to the team to decide what is appropriate.

1. A time series study of changes in the urban living and identifying distinctive features of evolving to the Smart living.
2. Study of theories of living interventions at the smart city level and modelling for study city.
3. The concept of accessibility in the Master Plan and its changes with respect to the increasing use of ICT in Smart Cities for smart living.
4. Changing the role of the hierarchy of service areas or watersheds in a Smart living in a Smart city as influenced by the increasing use of ICT.
5. Evolving structure of metropolitan urban agglomeration and changes required in a Smart city.
6. Evolving structure of cities in urban agglomeration and changes required in view of the increase in the use of ICT.
7. Change of spatial standards in a Smart metropolitan city.
8. Changes required in zonal policies and plans.
9. Study of Town and Country Planning legislation and suggest changes as per the special requirement of Smart living in smart cities
10. Change of role of community-based organizations (for example, Residential Associations in India) in a Smart city with an increase in the use of ICT.
11. Change of role of Ward Committee in a Smart city with an increase in the use of ICT.
12. Change of role of the Municipal Council in a Smart city with an increase in the use of ICT.
13. Change of role of the Planning system in a Smart city with an increase in the use of ICT.

Note: The scope of research can be further elaborated by the collaborating institutions but need not be uniform for all study cities. Each department of the university participating in this research programme shall incorporate relevant Smart Living Development features appropriate to the goals of each department. The coordinator of this project does not intend to dictate the direction of the research and have a diverse group of collaborating universities, and they should orient their study strictly based on academic goals of their department.

## 7 Study Cities

The study city will be selected as a study area by each of the collaborating universities independently, which will be the place the one-year and two-semester combined effort to conduct this research. Universities participating in this programme adopted different types of collaboration. Some universities used, their doctorate and post-doctorate students, while others used students at masters and first professional degree level. A post-doctoral student in the department can work on a narrow subject area in the study as individual work. While graduate and undergraduate students can work on design solutions for Smart Living, and Research institutions can charter their own strategic areas of research.

## 8 Project Details

One City will be selected as a study area by each of the collaborating universities independently, which will be the place the one-year and two-semester combined effort to conduct this research. Universities participating in this programme adopted different types of collaboration.

The project details of the study city are given in Table 1.

## 9 Way of Working the Programme

### *9.1 Integrating Smart Living Research with Academic Programmes*

This international collaborative research programme, with the participation of 8 countries and 16 study cities as tabulated above was conducted by many diverse university departments, research institutions and others as shown in the graph and table above.

### *9.2 Role of Students*

This international collaborative research programme is essentially meant for students since they are the future and being part of an internal academic programme of the university. We consider they are the main actor and shall be given an important role in this programme. Perhaps many that age group will live in the Smart Cities than their older faculty. Under the direction of faculty new concepts were introduced in the studio and empirical studies were conducted around these concepts.

### *9.3 Role of Faculty*

The faculty is the designer of the program within the framework of existing curricula in design studios and theory courses of each participating university.

1. The project duration is one academic year or two semesters.
2. They guide and monitor student work as usual as part of the academic programme.
3. They monitor students' input to the monthly progress report.
4. They rewrite the output of the project for a book to be published by an international Publisher giving due credit to their work.

**Table 1** Project details

| Authors, countries and institutions | | |
|---|---|---|
| Study city (Country) | Authors | Institutions |
| Hcng Kong (China) | Sujata S Govada, Timothy Rogers, Leon Cheng, Hermion Au, Hank Yuen | Institute or Sustainable Urbanisation, UDP International, Chinese University of Hong Kong |
| Bhopal (India) | Amit Chatterjee, Premjeet Dasgupta, Gaurav Vaidya<br>N. K. Paulose | School of Planning and Architecture, Bhopal, India |
| Delhi, Mumbai, Ahmadabad (India) | Ashok Kumar | School of Planning and Architecture, New Delhi, India |
| Dehradun (India) | Prabh Bedi | School of Planning and Architecture, New Delhi, India |
| Leh, Shimla, Dharamsala (India) | Anjali Sharma, Anuradha Cheturvedi | School of Planning and Architecture, New Delhi, India |
| Kozhikode (India) | T. M. Vinod Kumar | School of Planning and Architecture, New Delhi, India |
| | Bimal P, Namrtha Radharkrishnan, Krishnaveni K S | Department of Architecture and Planning, National institute of Technology Calicut, India |
| Surat (India) | Leena Garg, Krupesh Chauhan | SV National Institute of Technology, Surat |
| | Bhasker Bhatt, Kusumlata Khurana | Arvindbhai Patel Institute of Environmental Design, Anand |
| | Priyanka Kumar | Sarvajanik College of Engineering and Technology, Surat |
| Thiruvananthapuram (India) | Reghunatha Menon K. P.<br>Suresh Francis | Kerala State Remote Sensing and Environment Centre |
| Vijayawada (India) | S. Ramesh, Ayon K. Tarafdar, Krishna Kumar S. V, Ravi Kumar Reddy<br>Anand | School of Planing and Architecture, Vijayawada, India<br>Indian Police Service |

(continued)

**Table 1** (continued)

| Authors, countries and institutions | | |
|---|---|---|
| Study city (Country) | Authors | Institutions |
| Galilee Region (Israel) | Yifat Reuveni | Tel Aviv University, Israel |
| | Dr. Itai Beeri | University of Haifa, Israel |
| | Neomi Fnsch | |
| Lagos (Nigeria) | Femi Olokeusi, Femi Ola Aiyegbajeje, Samuel lyiola | Faculty of Social Sciences, University of Lagos, Nigeria |
| Dubai (United Arab Emirates) | Ashimita Karmarkar<br>Ashika Raoof | |
| Derby (United Kingdom) | Eleni Tracada | University of Derby, UK |
| | Eina Alatalo | University of Tampere, Finland |
| | Guglielmo Minervino | Mediterranea University of Reggio Calabria |
| | Joanna Solbe | Derby City Council |
| Cincinnati Cleveland, Pittsburgh, Atlanta, Tampa (United States of America) | Sudeshna Ghosh | Department of Geography and Regional Planning, Indiana |
| | | University of Pennsylvania |
| | Sweta Byahut | Department of Political Science, Auburn University |
| | Calvin Masilela | Department of Geography and Regional Planning, Indiana |
| | | University of Pennsylvania |
| St. Louis (United States of America) | David Sandel<br>Sharonica Harden<br>Robert Dillon<br>Francis Chmelir | |

## 9.4 Co-design and Co-production of Knowledge

This international collaborative research programme is founded on the principles of co-design and co-production of knowledge. In today's interconnected world, such collaboration is physically and intellectually possible—thanks to the Internet and ICTs. The collaborative aspect of the research programme will be actualized in the form of:

1. Co-design the programme with the partner academic institutions.
2. Co-production of knowledge through an interactive process of sharing, reviewing and finalizing research findings.

3. Within each partnering institution, co-design and co-production of knowledge can be implemented through design studio/laboratory work between faculty and students.

### *9.5 Research Output*

The key output of the "Smart living for smart cities" research programme will be a book edited by the coordinator Professor T.M. Vinod Kumar, to be published by Springer-Nature, an internationally reputed publisher in 2020.

## 10 Bulletin

During the conduct of research about 24 months through in 2018–19, three Bulletin has been used to communicate with the international teams of researchers. These Bulletin highlights study city profiles selected by the various study teams independently and introduces to the research network the research methodologies adopted, and the profiles of authors of the research output for the book, "Smart Living for Smart Cities". The Bulletin is jointly edited by a Bulletin team among authors, Namrata Radhakrishnan (India), Yifat Reuveni (Israel) and Sweta Byahut (USA).

The list of Bulletin produced with respective case studies is shown in Table 2.

A cover page of the Bulletin 3 and one sample page of the Bulletin is given below.

**Table 2** List of bulletins

| No | Country | City | Bulletin number/date |
|---|---|---|---|
| 1 | China | Hongkong | 1/30 June 2018 |
| 2 | India | Mumbai | 1/30 June 2018 |
| 3 | India | Delhi | 1/30 June 2018 |
| 4 | India | Ahmedabad | 1/30 June 2018 |
| 5 | India | Dehra Dun | 1/30 June 2018 |
| 6 | India | Kozhikode | 1/30 June 2018 |
| 7 | India | Kozhikode | 1/30 June 2018 |
| 8 | India | Shimla | 2/31 January 2019 |
| 9 | Israel | Galilee Region | 2/31 January 2019 |
| 10 | USA | St. Louis | 2/31 January 2019 |
| 12 | India | Bhopal | 3/30 June 2019 |
| 13 | UAE | Dubai | 3/30 June 2019 |
| 14 | India | Surat | 3/30 June 2019 |
| 15 | UK | Derby | 3/30 June 2019 |

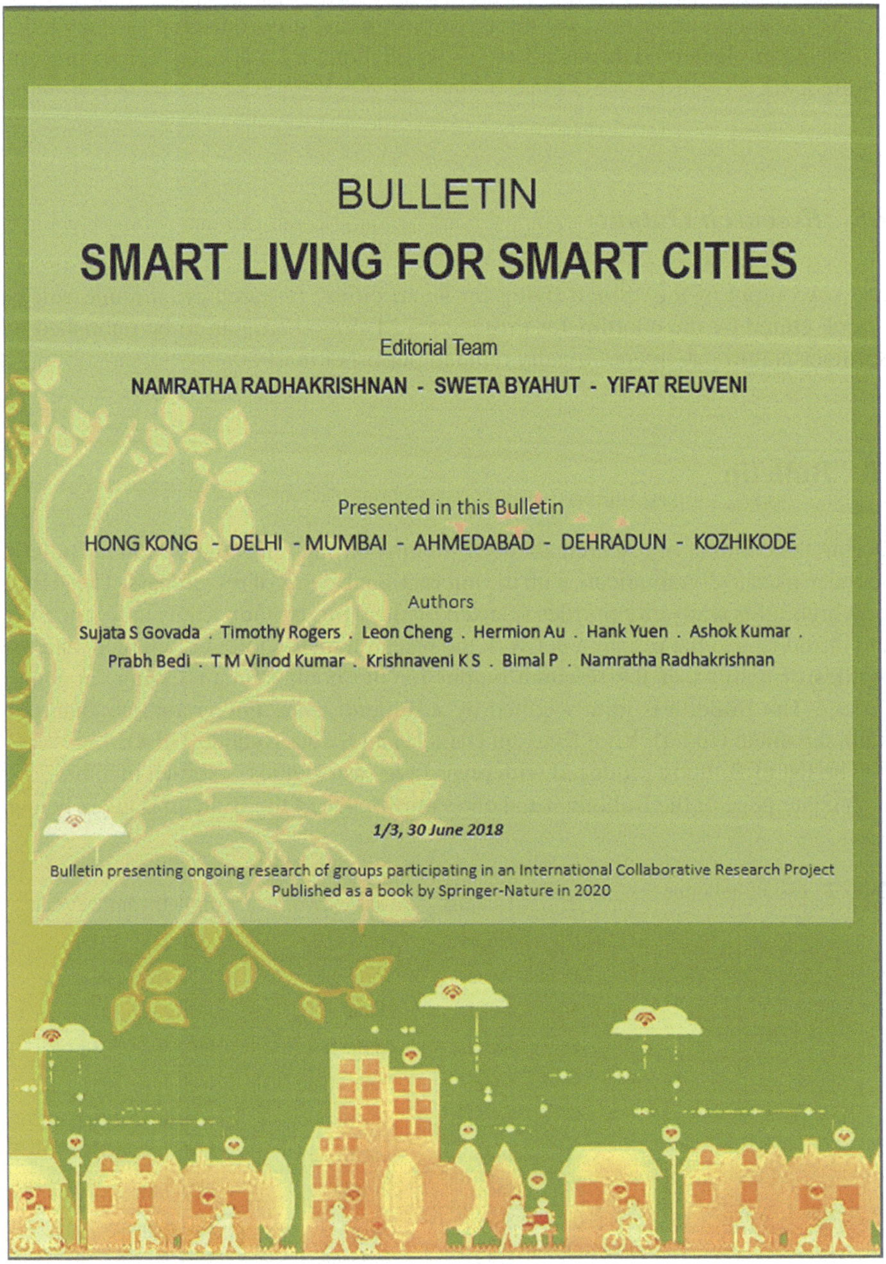


A page of Bulletin is as shown below (Figs. 3 and 4).

**Fig. 3** Kwung Tong promenade in Hong Kong. *Source* Energizing Kowloon East (2017)

**Fig. 4** Connecting green and blue assets to public and active transport. *Source* HK2030+ (2016)

# 11 Summary of Conclusions and City Case Studies

Team leaders of authors were requested to prepare a summary of conclusions of their city studies and the Editor compiled it and is given below.

## *11.1 Smart Living for Smart Cities*

### 11.1.1 Conclusion Smart Living

1. Smart living is one of six components of the smart city. Other five are smart people, smart mobility, smart economy, smart environment, and smart government. All smart living activities have also these six components integrated with smart city system.
2. A City provides with a complex space for all living. Wellbeing in a city for an individual is often satisfactory, sometime unsatisfactory and often threatened with city environment that is largely unfamiliar. No individual alone can control all the external environment of the city he is part of.
3. A deliberate design of smart living is required to be executed by the community. How to design such smart living is a challenge.
4. The smart city is a unique but complex mixture of cultural groups and varying spatially, ecological and cultural system. Because of the unique ecology of Smart Cities, a generalised model of the city is not possible. So also, a generalised smart solution for a smart city is not workable using ICT and IOT. The spatial and ecological structure of built space forming land use and transportation system also varies and as other ecological elements of the city. However, these ecological elements are ideal candidates for dealing with the smart city conceptualisation and for cyberspace interventions in real time using the Internet of Things and ICT.
5. The smart city is also having unique cultural characteristics. There are innumerable factors that influence the culture of a people in a smart city. The culture of the city is determined by the regularity of behaviour of inhabitants of smart cities. These factors determine the cultural attributes of smart cities. These cultural factors also have a bearing on smart city functioning and will influence smart living.
6. Study of the city is the study of this unique ecological and cultural system and smart living is the celebration of the unique ecological and cultural system that constitute the city by people. As the size of settlement increases so also the complexity. Then, a better way of study is the classification of human settlements into parts and then study it. Looking at the global cities Doxiadis divided the human settlement according to the population on a logarithmic scale of 15 parts/population size. These 15 levels can be used for the study of the ecological and cultural system of the settlement.

7. In shaping his settlements, man has always acted in obedience to five principles. As far as Doxiadis know, this has always been true, and he has not found any cases which prove the opposite. The first principle is the maximization of man's potential contacts with the elements of nature (such as water and trees), with other people, and with the works of man (such as buildings and roads). This, after all, amounts to an operational definition of personal human freedom. The second principle is the minimization of the effort required for the achievement of man's actual and potential contacts. The third principle is optimization of man's protective space, which means the selection of such a distance from other persons, animals, or objects that he can keep his contacts with them (first principle) without any kind of sensory or psychological discomfort. The fourth principle is optimization of the quality of man's relationship with his environment, which consists of nature, society, shells (buildings and houses of all sorts), and networks (ranging from roads to telecommunications). The fifth principle, man organizes his settlements to achieve an optimum synthesis of the other four principles, and this optimization is dependent on time and space, on actual conditions, and on man's ability to create a synthesis. Doxiadis concentrated mostly on the cultural system of human settlements while Patrick Geddes expanded it to ecological which he calls "nature and culture".
8. Geddes advocated the interdependence of culture and nature and merits in our approach to smart cities and smart living. He advocated the design approach in which one materially adopt people and livelihood to specific challenges and opportunities of a place they inhabit. This resulted according to him a marital expression of culture in spatial design in nature. Geddes' work was based on his fundamental principle of 'Place, Work and People' as he believed that geography; economics and anthropology were related, yielding a single chord of social life. He saw sociology as a quintessentially interdisciplinary subject that was essentially the science of man's interaction with a natural environment, and improved urban planning was one of the keys.
9. Geddes concept of life is one that connects culture with nature. He believes that nature conservation in cities, technology acts as a vehicle to meet human needs in a city that integrate ecological and social processes to create a healthy environment. The current competitive outlook will give way to cooperative outlook at the regional, international and global scale. He was foreseeing regional economy based on renewable energy.
10. Urban space according to Lefebvre is a social construct. He extends the concept of Geddes further and what Doxiadis implied it in his work. He said, once you repeat social activities in the same place, it converts itself to a cultural construct. It is created through social relations that Lefebvre characterised as a triad of spaces: (1) spatial practice, (2) representations of space, and (3) spaces of representation.
11. Advaita (Non-dual) Vedantic (end of Veda) view of living is very briefly discussed by presenting, one Upanishad attributed to Mandukya and Mandukya Karika which elaborate the Mandukya Upanishad by Gaudapada wrote in the seventh century AD, in a commentary known as Mandukya Karika. The Karika

is based on Mandukya Upanishad that examines various states of human being to determine the reality of existence as he experiences it which opens a way of living through awareness which Gaudapada calls as Asparsa Yoga without any dharmic codes of conduct. This awareness is also central to smart city concept as discussed earlier and Gaudapada put forward the concept of Asparsa Yoga in the concept of self-awareness to implement in daily living.

12. The Buddhist way of life accepts the cultural and ecological system in a city that envelops all. It understands the influence of this system on human existence and prescribes in concrete terms understandable by a common man how to deal with it and conduct living. Buddhism cannot be there without the Sangha, the modern-day smart community and cannot be there without the Governing codes and Democratic norms of the much appreciated and celebrated Republics in the Buddhist era.
13. The design of smart living can be worked out based on the ecological and cultural system in a city. In a theocratic country probably, religion may be the focus of cultural and ecological system that affect the life. For example, Judaism in Israel, Christianity in the Vatican, Buddhist way of life in Tibet in China and Islam in Saudi Arabia can be the respective focus of the cultural and ecological system of cities in Israel, Tibet in China, Vatican and Saudi Arabia respectively. In a secular country there can be many foci and each federal state can follow that focus based on the statistical indicator of the relatively high performing aspect of life such as for example health sector which has thrived over few centuries in the Kerala state of India. Under secular constitutions like India, no one religion is there to shape the cultural and ecological system. Many religious ways of life coexist side by side in the same city. Many laws are framed for different religion for practice as per the constitution of the country. In such countries the descriptive statistical analysis of achievement variables can be a guide.
14. The smart part explained with respect to ekistics scale 1–3 may become more complex for ekistics scale 4–15 but must be designed with respect to the primary focus of the cultural and ecological system. In secular countries like India living may be focused on some other aspect since in a secular country if smart living is focussed on majority religion, it is not politically correct since all religions are of equal status, irrespective of the relative size of the population. In a cultural system, all manifestations and patterns of living are equally valuable and no ranking. In this case we must shift the primary focus to something else based on the local empirical reality. Again, this primary focus may differ from regions to regions within an urban agglomeration or city in a secular country. This brings about a diverse and exotic smart living. Whatever the focus the domain study shall proceed with the design for smart living. The domain can be a religious practice, constitution, or certain aspect of the culture values most as demonstrated by statistics say health along with consequent longevity. This is presented in Chapter "Geospatial Intelligence for Smart Living–Case of New Delhi" for Kozhikode Metropolitan Region.

15. Smart city is a knowledge-based city that develops extraordinary capabilities to be self-aware, how it functions 24 h and 7 days a week and communicate, selectively, in real time knowledge to citizen end users for a satisfactory way of life with easy public delivery of services, comfortable mobility, conserve energy, environment and other natural resources, and create energetic face to face communities and a vibrant urban economy even at a time there is National economic downturns. Smart Living in Smart Cities is that gives people the opportunity to benefit from new ways of living. It involves original and innovative solutions aimed at making life more efficient, more controllable, economical, productive, integrated and sustainable. Smart Living" enhances the overall living experience of the people across age groups and demographics. The Smart Living focus on improving an individual's ability to interact with electronic services, and on improving general wellbeing. The aim is to create a safer, more secure, accessible, and happy society.
16. A synthesis of many considerations mainly strategic together leads to the design of smart living for smart cities. They are discussed below very briefly. They are enumerated below.

    (a) Design for the development of an ecosystem for innovation in smart living.
    (b) Spatial Design for Smart Living.
    (c) Design of Smart Community.
    (d) Design of Smart GIS for Smart Living.
    (e) Design of GIS-based Spatial Decision Support System (SDSS)for E-Democracy and E-Governance for smart living.
    (f) Design of GIS-based Zonal Plan Design including Hybrid Form-Based Codes for smart living.
    (g) Design the ICT system for use in all smart living activities.
    (h) Design of IO system for all smart living activities.
    (i) Design modifications to institutional arrangements for smart living implementation.

17. Having discussed various aspects of the smart part of living, let us discuss the strategy for applying it to the ecosystem for innovations that leads towards a smart city is given below.

    (a) Invigorating the ecosystem for innovation to support the development, commercialization, scaling up and export of smart solutions. A smart solution to smart living can be both hardware and software in terms of smartphone apps as well as innovating smart protocol the community design. This means conceptualise, design, develop and deploy smart living devices for ecological and targeted to living by species, sex and age in collaboration with the smart community where mass scale adoption is visualised locally and the possibility of global demand.

(b) Engaging the public and business and co-creating smart solutions with stakeholders to better meet the needs and fulfil the aspirations of the people: This means establishing mutually beneficial linkages with smart community and business and universities for field testing and adoption and opening business opportunities.
(c) Cultivating a culture of embracing changes and collaboration for smart city development: this calls for continuing education as well as skills the smart community.
(d) Integrating with city planning; This may involve changing the planning legislation or amending it. This also calls for educating the Planners on Smart Living.
(e) Establishing a sound digital framework to support further improvement to the operations of urban infrastructures and provision of services; There shall be a program to augment backbone infrastructure of the existing ICT and IOT system in the annual budget of local bodies.
(f) Providing support to reduce the digital divide and ensure inclusiveness of Smart City development. The policy of free provision of smart devices to the economically weaker households. Opening bank credit at concessional rate for smart device procurement by the smart community.
(g) Equipping students and workforce with the knowledge and skills to seize opportunities in a Smart City: Skill development of students in emerging ecology based smart devices and its application. Creating an ecosystem for innovative start-ups.
(h) Developing an innovative strategy for procuring smart solutions: Smart Community shall form a close alliance with local technological institutions. This alliance shall lead to discuss problems and design and implement innovative solutions from these technical institutions.
(i) Improving preparedness, prevention, and mitigation of adverse events. Smart Community shall have a dedicated group of members to achieve this task year-round.

18. Spatial Smart Living Strategy is as follows

(a) Spatial means regional which happen to be a programming region of the metropolitan committee. There are no programming regions for megacities and megacities in India and many countries. For example, Kozhikode Metropolitan Region is the constitutional body eligible for consolidated funds of union and State Government of Kerala and an elected body for its development. Although it is in the constitution it is not established yet in many states like Kerala.
(b) There is a metropolitan committee to govern it as per the Town and Country Planning Act 2016 of Kerala (under revision). One of the goals of this committee shall be smart living for all and there shall be a budget to meet it.

(c) It has constitutional functions to perform as per Act 2016 with definite membership. Membership may be expanded to include industry and academia which is not there.
(d) The regional strategy of smart living means settlement, linkages and infrastructure strategy. Infrastructure includes ICT and IOT infrastructure for smart living.
(e) Regional means consideration of all natural (ecological) and human resources and working out the resource-conserving strategy for smart living.
(f) Regional means it is a polarized region with emerging settlement hierarchy and functional specialization which provides a spatial matrix to design smart living Strategy to increase metropolitan wellbeing and clearly point towards sub-regional specific scope of work.
(g) Spatial smart living strategy to avoid sub-regional metropolitan area disparity and make metropolitan wellbeing inclusive of poor and rich and middle class.
(h) Strategy to increase metropolitan employment opportunity as related to smart living activities for locals and immigrants of all categories. Smart living is a generator of employment.
(i) Strategy for use of natural and human resources and including strategy for import of these resources if not available locally.
(j) Strategy for development of related institutions for a well-being.
(k) Smart Living with Smart Economy, smart people, smart energy, smart environment, smart mobility, and smart governance in smart city framework.
(l) Smart wellbeing and health initiative by deploying and using required ICT, and IOT.
(m) Design for empowering households and community for smart living.

19. Smart Community Strategy is as follows

(a) A smart community is identified as Residential Welfare Association (RWA). A neighbourhood as conceived by planer in not easily spatially discernible but RWA is a registered entity under society act involved with the welfare of the local community.
(b) RWA is below Ward Committee area, Panchayat, Towns and Metropolis. Those spatial unit above RWA are generally constitutional body but RWA is not. The constitutional body must draw annual plans as per Governmental protocol and spend their share of tax money collected by State and Union Government, as well as local bodies as per the directive of Financial Commissions and, are to elect members by regular election in periodic intervals under the constitutional body the Election Commission. RWA is also accountable to themselves under society act and the election is required to be conducted as per their registration document.

(c) RWA consists of about 100 households mostly known to each other. In a Kozhikode Metropolitan Region, there may be around 20,000 plus RWA with a total population above 2 million.
(d) Metropolitan Area as per constitution is required to prepare hierarchy of plans such as perspective/vision 20 years, midterm 5 years and annual plan tied up to the annual budget provision of the Metropolitan Area. This plan shall focus on smart living and include a component implemented by RWAs. The methodology for the plan will be like what is presented in Chapter "Role of Smart Land Scapc Architecture in Smart Development of the UAE Case Study-Dubai" for Kozhikode Metropolitan Area.
(e) Converting the RWA community of 400–500 people in a walkable area even for children is to be executed in a systematic way. The action program shall be drawn by RWA itself.
(f) RWA shall be subjected to continuous training in skilling to meet the local needs as dictated by metropolitan plans for smart living execution. Households in Kerala are fully computer literate and have been declared as such. This computer literacy makes them familiar to operate computers and the internet and send emails etc. This Skilling shall be expanded to Geographic Information System Data Generation of RWA area GIS by using public domain GIS software and use a Spatial Decision Support System for smart living as discussed in Chapter "Geospatial Intelligence for Smart Living–Case of New Delhi". They may be trained to update RWA GIS once in every 3 months with new features and attributes as related to smart living.
(g) In addition to GIS, they shall be educated as end users of various IOT system for smart living and assistive technologies for smart living by relevant experts from the Metropolitan Region or outside.
(h) For continuous training for skills in smart living planning, execution and management RWA shall establish relevant linkages with Universities, Colleges and Vocational Schools adjacent to RWA.
(i) RWA shall be trained to conduct E-Democracy at RWA area.
(j) RWA shall assist the Government to implement legislation and acts both from central Government and State Government. This is discussed in Chapter "Geospatial Intelligence for Smart Living–Case of New Delhi".
(k) Spatial Decision Support system shall be designed over GIS as a collaborative exercise of RWA and relevant smart living issues as discussed in Chapter "Geospatial Intelligence for Smart Living–Case of New Delhi" with the linkage already established with the relevant institutions.
(l) There shall be a set of E-Governance SDSS and E-Democracy SDSS as discussed in Chapter "Geospatial Intelligence for Smart Living–Case of New Delhi" which RWA use daily for smart living.
(m) For Government support for smart living planning, execution and management RWA shall establish relevant linkages with State and Central Government Offices in the Metropolitan Area. This will ensure full Government support for smart community action.

(n) RWA shall assess from Metropolitan Plan their role on a 24 h 7 days basis in implementing Smart living.
(o) Smart Community shall be organised to implement these protocols as discussed under section smart community in Chapter "Geospatial Intelligence for Smart Living–Case of New Delhi".
(p) RWA shall organise free hosting for students of technical universities in their selected house who come forward to develop free smartphone apps for e-Democracy and e-Governance for smart living.
(q) RWA shall organise free hosting for students of technical universities in their selected house who come forward to help as free volunteers for geriatric care and palliative care. For e-Democracy and E-Governance for smart living.
(r) RWA shall develop in collaboration with cell phone operators the backbone of ICT and IOT system in their areas.
(s) RWA shall organise free hosting for students of technical universities in their selected house who come forward to design and develop a public realm for smart living.
(t) RWA shall identify selected houses for senior citizen congregation and activities for smart living.
(u) RWA shall identify sites for locating and siting IOTs for specific use in conjunction with smart living with the help of relevant institutions in the metropolitan areas.
(v) RWA shall identify server location for smart living cloud computing in relevant institutions.

20. Major questions of Smart Living in Cities, Metropolises, Megacities and Meta cities.

(a) How to realize existing institutional capacity to implement Smart Living for Smart Cities program in Metropolitan Areas and design ways and means to plan and implement. I do not recommend special purpose vehicles which may weaken existing constitutional bodies.
(b) How to use the Constitutional power and responsibility vested in the Metropolitan Planning Committee to plan, design, implement and govern for a smart living for smart cities.
(c) What are the existing provisions in terms of infrastructure for smart living in study areas, gaps, potential and SWOT.
(d) What constitutes place making for a design for need-based Physical, Mental, Social and Ecological Health for smart living in the Metropolitan area?
(e) What constitutes place making for a design for cultural needs for smart living in the Metropolitan area?
(f) What constitutes place making for a design for a sharing community production and services needs for smart living in the Metropolitan area?

(g) How to prepare and use various types of zonal codes designed largely by the community including hybrid Form-Based Codes and design modules to implement smart living strategies. This Chapter will answer it.
(h) How to design and use ICT and IOT system for a smart living.
(i) How to design E-Democracy for smart living including a spatial decision support system.
(j) How to design E-Governance in Smart Living including spatial decision support system.

21. Planning for Smart Living in a Smart City essentially is answering community how the following tasks may be performed in the plan for smart living.

(a) What is the concrete term are the Concepts of Smart Cities, Smart Living, Smart Communities, Smart Living E-Governance, Smart Living E-Democracy for the smart community for the study area?
(b) What are the examples of smart living which we can study as replicable good examples of smart living in smart cities programs from India and abroad?
(c) What should be the Aims of regional or community study achievable, what are its objectives, what methodology to use, Service Area Delineation, Area Profile with respect to the smart living, Mapping and secondary data collection.
(d) How to assess existing Living Situation Physical, Mental and Social Health, Existing land use standards as related to living, Public Realm for Physical, Social and Mental Health of all age group, Existing Public Health Infrastructure, Existing Access and Mobility to the public realm and health infrastructure, Existing Environmental Health.
(e) How to assess and specify problems and issues of living, Physical, Mental and Social Health, Smart City Policies and standards, Health Policies and Standards, SWOT and Gaps.
(f) Assessment of Living in Study Metro- from domain foci example health statistics, land use studies, environment studies, mobility studies of public realm and comparison with selected metros with the high quality of life.
(g) What Geospatial analysis of study area using RS and GIS for Smart Living of ecology and ecosystem of the area, Geospatial analysis of Environmental Governance using Remote Sensing and GIS and legislative norms and direction spatial Decision criteria such as CRZ etc. can be used for the above objectives.
(h) How to conduct a Study of smart living and related quality of life indicators.
(i) What are the Existing Land use needs and provide for human needs by age groups and income for a smart living?
(j) How to the assessment of ecological needs and provide for human needs for smart living.

(k) What are the Institutional set up for Urban Planning, Urban Development and Urban Management in the study area, Constitutional responsibility of the Metropolitan Planning Committee, Municipal Corporation, Municipality and Urban Panchayat to deliver smart living? Financial Capability for urban development for smart living, Capacities and gaps in executing Smart Living Projects in the study area, SWOT analysis of Institutions for smart living projects and suggested remedies.
(l) How to briefly Summarise Assessment of Issues and Gaps.
(m) How to conduct a SWOT analysis and Vision of Smart Living.
(n) How to design Land suitability analysis using RS and GIS for the ecological needs of the city.
(o) How to conduct Identification of location for public realm for local needs.
(p) How to assess the Economic Opportunities in the Public Realm, Existing Public Realm Governance.
(q) How to decide on Design standards and modules for the smart living of the public realm.
(r) How to conduct Designation of Euclidian, Performance and Incentive and Form-Based Codes for Smart Living.
(s) How to design E-Governance for Smart Living based on existing legislation as related to living.
(t) How to design E Smart Living Democracy, Principles, Concept, Methodology, Case studies, Proposals for Smart Living for executing smart living in the area.
(u) How to design IOT and ICT System for Smart Living using existing infrastructure.
(v) How to frame Proposals: Urban Land Use Standard for Smart Living, Zoning standards for smart living, Hybrid Form-Based Codes and Smart Living Standards.
(w) How to frame Proposals: Smart Living Design Modules for Hierarchy of settlements.
(x) How to frame Proposals based on Multicriteria Spatial Decision Support System for Smart Living spaces.
(y) How to frame Proposal: Identification of living spaces and creation by urban land management for humans by age group and design criteria, Identification of living spaces for other living by type and design criteria.
(z) How to frame ICT and IOT for Smart Living, IOT for smart living skills, IOT groups for Smart Living skills, IOT scenes for Smart Living skills.
(aa) How to frame ICT proposals and its use for smart living, Infrastructure proposals for ICT and IOT.
(bb) How to create Smart Living Communities.
(cc) How to design Smart Community for smart living management, Legal structure, Financing Strategies, Bylaws.
(dd) How to create Smart Community design by smart living activities and registration as a society with specific site-specific objectives of smart living.

22. Planning, managing and development of Smart living for smart cities is based on some important principles of smart cities.

    (a) Smart living involves sharing freely human resources for smart living. All living activities are fully supported and shared by all involved locally without monetising this sharing. It is the generator of strong community feeling locally and making people aware of all happenings around them and find effective roles to face it.
    (b) Smart Living shall be based on the demand of local culture, local ecosystem and dominant values of the community such as religion, health etc.
    (c) Smart Living strengthens and empowers local community by use of ICT, IOT, E-Democracy and E-Governance designed as per local needs by local people and institutions.
    (d) Smart living is based on a holistic and comprehensive view of life.
    (e) The marginal cost of sharing is kept near zero to make smart living sustainable.
    (f) Smart Living is essential to work with the local smart community.
    (g) Smart living is possible only if all local institution of higher education helps the local community in their continuing education and skilling to manage smart living as well as local planning.
    (h) Smart living requires sharing facility and services of local institutions to help create a smart living.
    (i) Smart community assists in Governance as related to smart living by using smartphone based on E-Governance Applications for a smart living using the help of local institutions of higher learning.
    (j) Smart Community assist activating smart living by use of use smartphone-based E-Democracy of Applications for a smart living using the help of local institutions of higher learning.
    (k) IOT system for smart living shall be developed and managed by the smart community with the help of local institutions of higher learning.
    (l) Server for smart living apps will be in an institution of higher learning and research locally.
    (m) Wherever possible data mining and artificial intelligence will be applied for the speedy delivery of smart living.

## *11.2 "Urban Transformation for Sustainable Growth and Smart Living: The Case of the Atlanta BeltLine"*

The chapter on "Urban Transformation for Sustainable Growth and Smart Living: The Case of the Atlanta BeltLine" undertakes a comprehensive literature-based assessment of the Atlanta BeltLine project being implemented since 2008. This study brings multiple aspects together in a single case study across four smart living perspectives, namely, urban health and wellbeing, smart public transportation, economic development and brownfields clean-up, and managing affordable housing during neighbourhood change. The BeltLine project in Atlanta combines elements of urban greenway planning, parks and recreation, pedestrian and bike friendly trails, public transit, and other facilities connecting about 45 urban neighbourhoods and estimated to cost about $4.8 Billion. The project is expected to attract about $20 billion in investments over 25 years to bring about significant new businesses, jobs, and other economic development opportunities in the redeveloped areas, and expected to create 30,000 permanent jobs. It is already bringing about transformative urban regeneration in those neighbourhoods, with large scale economic investments and growth in previously neglected areas and brownfields.

The authors have used extensive literature review and undertake GIS analysis using U.S. Census Bureau datasets, which provides a snapshot analysis of socio-economic changes happening in the neighbourhoods surrounding the BeltLine over time. They compared data for the year 2000 (before the implementation of the project) and year 2017 (the most recent time period for which data is available). This captures the effect of BeltLine project implementation and generates some key insights on socio-economic changes that have been witnessed in these neighbourhoods across four indicators: income and educational level, the housing market, racial distribution, and poverty level.

The authors find that while the BeltLine has brought about considerable growth and investment and created city level green spaces and trails, the process has been accompanied by gentrification and displacement that has disproportionately affected the racial minorities and poor neighbourhoods. They find that often the positive values of such large scale and transformative urban regeneration and greenway programs are not much acknowledged. Considerable economic, cultural, and experiential outcomes are just as important as health and transportation outcomes.

However, during the BeltLine implementation over the past decade, priorities have changed to focus more on recreational trails and urban parks development, while the crucial light rail transit component and affordable housing issues have been side-lined. While the BeltLine has resulted in considerable growth and investments, and created city level amenities such as green spaces and trails, this redevelopment process has been accompanied by gentrification and displacement, disproportionately affecting the racial minorities and poor neighbourhoods, and current policies and programs are not able to address these issues adequately. Economic cost benefit analysis of smart living ideas caters to a few specific groups of people, which lead to questions about equitable distribution, social justice, and the rights to the city. The larger question is- how can the different urban sustainable development goals be balanced in these transformative urban projects?

## *11.3 Exploring Liveability in the Era of Smart City: A Case of Bhopal*

Bhopal is strategically located in Central India and the Capital city of the State (Province) of Madhya Pradesh. The city is also known for its knowledge hubs, culturally rich communities, and business hub. Out of the 100 cities chosen under the Smart City Mission by the Government of India, Bhopal has selected in the first round of city selection. The Smart City Mission envisions transforming the 'city of lakes', Bhopal into a future-ready Smart City with enhanced liveability, workability, and sustainability. The present study focused on accessing the liveability condition based on following the three parameters: (i) Smart transportation, (ii) influence of smart living on urban informality and (iii) good governance, especially e-Governance and citizen participation.

Public Transport modes available in Bhopal included BRTS, Standard Buses, Intermediate Public Transport (Mini-Buses), and Para Transit (Vans and Shared Autos). The proposed Bhopal metro rail project will include a total of six lines in a phased development. As part of smart city initiatives, Bhopal Smart City Development Corporation Limited has taken up projects for Smart traffic management, which includes Smart Road initiatives, Smart Public Bike Sharing, Smart Parking Management, and allows citizens to make informed travel choices through Integrated Traffic Management System. A liveability assessment of Bhopal based on the Ministry of Housing and Urban Development formulated 12 indicators from the mobility sector show that the city is slowly progressing towards better liveability. Also, there is a decreasing trend observed in road accidents, loss of life, and injured people, which indicates that the travel demand management measures adopted in the past one decade in terms of improvement in public transport, creation of pedestrian and bicycling infrastructure, and introduction of the smart traffic management system.

It emerged from the detailed field investigations that the use of smart technologies is changing the business practices and communication trends for street vendors, and with time, more people will join this flow. Street Vending Act gave legal identity to them, and the latest modernization can encourage them to explore new avenues for improvement of their trade and incomes, such as Local Vending App, Smart Vending Zones, Vendors Insurance and Credit system etc. These initiatives will change the character of informal employment without converting it into a traditional formal sector.

The twenty-first Century will be expected as a new era for urban informality. Smart governance will resolve development conflicts between informal and formal housing systems by connecting aspirations and promoting interdependence for the collective growth of people from various social strata. At the same time, with a considerable share of the economy, the informal sector (employment) will be treated as a primary resource to explore the various dimensions of economic development, which is also called Human Capital. Smart transportation will be the backbone of upcoming development. Smart transport infrastructure and smart transport services will attain fast, safe, and economical movement of passengers and commodities over the years. Demand oriented skill development, entrepreneurship promotion, and systematic approach for the use of latest technologies for various socio-economic groups, effective coordination in stakeholder's consultation, the role of information and communication technologies, e-governance, and community participation in the planning process and financing strategy development are the game-changer in the coming future. Collectively, the above said efforts would make living smart for people under urban informality.

An attempt also made to investigate the socio-economic and technological factors that are most likely to be associated with ICT maturity. The usage levels of the Bhopal Plus app and the Mayor Express app were mapped for each ward of BMC. It was observed that penetration of the app is very low in those wards dominated by high proportions of low-income households in the Bhopal city. The use of smart parking and smart payment process for municipal charges is still not widespread. Most vendors have their individual savings bank account, but a lot of them have still not linked their accounts with Aadhar card and mobile phone, and that is why the share of on-line fund transactions is quite less. After linking it with the socio-economic profile of wards and the overall usage of digital governance apps, it is seen that the apps are not probably reaching out to the lower-income groups effectively, pointing towards a lack of inclusivity. Ownership of digital assets is quite high among slums and relatively low-income people, but still, the usage of Mayor Express and Bhopal plus apps are quite low in those wards dominated by low-income people. Thus, a case can be made that the digital urban governance innovations lack inclusivity in Bhopal.

## *11.4 Spatial Intelligence for Smart Living: Case of New Delhi*

The recommendations emerging from the chapter is given below.

1. The initiatives to promote citizen participation should adhere to a definitive and organised plan, which should be based on insights gained from geospatial data to ensure the plans are effective in realising their goals as well as efficiently doing so. Online citizen engagement, crowdsourcing are some of the known techniques, already implemented by many private companies, in which organized data, in combination with statistical analysis would bring about a qualitative dimension to the information and enhance the quality of the plan implementation.
2. Current geospatial models do not have well defined targets, functionalities and properly organised data, though they make available lots of useful mapping and query-based features. Fixing the lacunae will go a long way to ensuring the usefulness of leveraging ICT in projects. Since most of the technology is obtained from private players, the government users of geospatial data could ensure a research-oriented atmosphere for collaboration with private companies as well as academic researchers in extending the cause of geospatial planning.

3. The data collection process should ensure consent among the stakeholders and be robust enough to handle exigencies like unforeseen events which disrupt the collection process leading to gaps and inaccuracies in the collected data. Statistical analysis should be used to identify correlations.
4. A mechanism for modification of existing GMP of Delhi should be implemented, with each proposed change being scrutinized before adoption. In addition, environmental data needs to be incorporated into the GMP which is currently absent. Data updating and analysis beyond basic map visualisation should also be included in this portal.
5. Databases like DSSDI, ENVIS, USIS should be integrated with each other, and full scope of technologies like GIS like visualisation, delineation, query-based analysis should be exploited to lead to integrated and context aware decision making.
6. Delhi Geoportal should integrate a direct communication mechanism to cater to location specific conflict resolution.
7. Coordination between different government departments, databases should be ensured to realise the vision of holistic spatial planning of Delhi. Awareness campaigns aimed at the residents of Delhi should be conducted to ensure public co-operation to the data collection process.
8. The DSSDI needs to create a public platform for data management and ensure their data has a public outreach. It also needs to integrate a spatial dimension to grievance redressal and participatory public action. It should be able to reach a stage where the public can regard it as a necessity in their lives. It should be accessible to people in a language and visual ease which takes cognizance of their socio-cultural inclinations and should inspire the people to learn and understand more about the spatial components of their cities.
9. There should be an enhanced level of transparency regarding the sources, models and mechanisms for geospatial interventions in the Smart City Project in Delhi. Public participation in project formulation and its subsequent geospatial component needs to be enhanced and imbibed in the public imagination. Spatial Planning should not limit itself to a city level agenda and should transform its relevance to a national level development agenda through vertical and horizontal integration across the nation.

### 11.4.1 The Conclusions Are Given Herewith

The interface of smart living in New Delhi is realized through the Smart City Project. GSDL or Geospatial Delhi Limited is playing an important role in the conceptualization, development and implementation of spatial data services which are indispensable when it comes to smart city projects. For this reason, much of the DSSDI datasets and infrastructure is being modified and undergoing a process of augmentation and updating to assist in smart living projects. The Electric Vehicle Project, one of the most significant drivers of mobile smart living in New Delhi has included aspects of GIS assessment and evaluation in their framework. Bicycle routing, utility network upgradation and sustainable development initiatives, all require geospatial interventions in some capacity. In India, geospatial intelligence for a smart living is a relatively new concept. It is in the nascent stage of its integration and assimilation into the public and policy imagination. Although much of the spatial data for NCT of Delhi has been accumulated under the DSSDI, the same cannot be testified for other cities in the country. In such cases, the data infrastructure is so wanting that a complete conceptualization of informed decision making, and planning infrastructure needs to be brought about.

Spatial data is not just about discrete elements which are visible on maps. It is intricately related to people who are the forces and drivers behind the manifestation of spatial forms and patterns. The current infrastructure tries to enhance institutional delivery but fails to anticipate a realm for public engagement and people's participation in decision making. In fact, the geospatial infrastructures, if utilized to its fullest potential would create a vibrant series of democratic people's policies, reducing much of the paperwork, unaccountability, opaqueness and bureaucratic inefficiencies that often dissociate planning from its on-ground implementation. There is, therefore, a need for spatial literacy, beyond the purview of its use for little day-to-day activities, in the form of a dimension as important and portentous as time. Geospatial Intelligence is the culmination of all interactions between people, their environment and technology and it is to imbibe this sensitivity towards space and its **implications that we need delve deeper into its potential for revolutionizing smart living in smart cities**.

## *11.5 Smart Landscape Architecture for Smart Living: Dubai*

Development of Cities is often judged by the economic condition, commercial activities and building forms. The emirates cities of UAE are known for a high standard of lifestyle, safety, security and beauty. The city has expanded from a small trade centre to a large commercial hub.

The extreme weather of UAE gave rise to clusters of housing settlements with a very limited number of window openings that restricts direct sunlight and hot air within the habitable areas. Light and air could pass through the small courtyards, corridors, sikkas, wind towers and vertical slits within the roofs of the buildings that could reduce the harshness of the climate and provide diffused light and cool air to the buildings.

With the increase in urbanization modern buildings with large glass facades, balconies, openings developed during the expansion of the city like any other parts of the world. The modern buildings with utmost thermal comfort restricted people's movement in the outdoor open spaces.

The advent of technology and digitization on top of the indoor comfort restricted people's movement more towards outdoor spaces and public realm and slowly the region started losing the use of the outdoor spaces.

It is only during a span of the last few decades the region got transformed primarily to a cluster of working cities and the traditional public spaces that were underutilized again started coming back to life. The vacant sikkas, courtyards, plazas got transformed into spaces of assembly, leisure, enjoyment and breathing space with a new theme within the built concrete forest of the modern city development.

Slowly these public spaces, parks and playgrounds became an integral part of the city design and practice of landscape design became an important part of urban planning and design of UAE.

The recent drive to integrate all communities with the help of the public realm by using land scape features and smart technologies in allocating various active spaces is a move towards a unique healthy urban development.

UAE has proved to be a trendsetter in terms of embracing digital technology by adopting digital governance approaches, utilizing information and communication technologies for the development of the region to establish itself as the "smartest" and the "happiest" cities in the world.

The region is also planning to expand its utilization of advanced digital means for sustainable growth and development with better governance and public wellbeing as a part of the leadership approach and collaborative governance style.

As a part of the strategy, the initiative is being taken in transforming and integrating municipal services that include creation and connection of integrated solutions for public open spaces such as parks, beaches and public realm within and across the emirates as the major urban planning initiatives.

Challenges of UAE:

- High temperature

UAE's climatic conditions such as high temperatures, evaporation, relative humidity and low average rainfall are the major challenges for the sustainable development of green open spaces and landscaped areas.

Climatic changes and frequent drought periods have led to a decline and degradation in natural pastures, resulting in increased numbers of grazing animals per unit area, which in turn led to the disappearance of favourite grazing plants and the spread of least palatable plants.

- Saline ground water

In UAE seawater intrusion is also a main cause of water salinity in the aquifer, especially near the coast. It is estimated that about 10 MHa of agricultural land is vanished annually due to salinization, of which about 1.5 MHa is irrigated land. Furthermore, the saline areas are increasing at a rate of 10% annually for various reasons. The groundwater salinity in the northern part of the UAE has increased abruptly and so as the central part of UAE to some extent.

- Lack of rainfall and fresh water

UAE has limited freshwater resources with average annual rainfall ranges from 80 and 140 mm the groundwater potential of both deep and shallow aquifers is about

757.6 $km^3$ while fresh water is less than 7.5%. The annual groundwater extraction is about 2668 million $m^3$ compared to recharge i.e. 350 million $m^3$

Smart Landscape architecture practice in UAE:

Smart landscape architecture is a practice that enhances the greenery by taking into consideration the conservation of resources. This practice involves the humankind from all directions and of all ages to a common space that is used for meeting, leisure and entertainment. It enhances and beautifies an open space through smart features and minimum maintenance.

For places like UAE, smart land scape architecture has become a special platform for urban development.

With smart use of resources, smart selection of the flora, smart technologies to expand the green open spaces and smart way of preserving the existing green spaces, smart ways to provide safe and secured green open spaces and smart usage of the open spaces with innovative activities. Some of the smart landscape practices of UAE are as below:

- Sustaining Water Scarcity and enhancing Green Growth.
- Promoting Sustainable Landscaping.
- Expansion of green with low water consumption.
- Forming regulations to create and link the open spaces and public realm.
- Use of Bright Colours, shapes and materials as street furniture and urban features.
- Use of lighting to create an enhanced look for the open spaces.
- Use of innovative street Furniture for people's comfort and beauty.
- Use of digital elements such as smart palm trees, digital signage boards, devices to enhance the utility of the open spaces.
- Adding adequate parking, provisions, waste management provisions and shower spaces/refreshing areas to the open spaces for the bikers and pedestrian to cope up with the climatic challenges.
- Smart entertainment system- Open air cinema, activities to add value to the open spaces.
- Use of intelligent irrigation system to control water consumption.
- Distribution of green space to various levels such as roof top, balconies, walls etc.
- Use of drones to monitor the conditions of the green open spaces.

Preservation and conservation of Green open spaces in UAE:

As per the practice in UAE parks are open spaces are created in limited numbers but larger in size. All the parks in the UAE are planned to include multiple activities and facilities with ample landscape features. The parks are also facilitated with ample parking and safety security arrangements that can support large scale events arranged within the parks and open spaces. Limited large open spaces and parks are easy to maintain with the installation of smart gadgets and few control/monitoring stations.

The developer of the community maintains the parks, open spaces and public realm. The maintenance charges are paid by the community residents in terms of service charges, hence there is an involvement of the government, developers and residents in the creation and conservation of parks and open spaces of UAE. The property developer's role is to select the smart technologies for their development in such a way that the development becomes sustainable, energy saving and attractive for the users. A smart selection of the elements and technologies will also add value to the property with a great chance for a high return on the invested amount.

Performance monitoring of the parks and open spaces commences continuously through various ways and means with the judgement from residents, visitors and the users of the facilities. The feedback from various sources is considered for the improvisation of the areas and the facilities within the spaces. Smart gadgets to obtain user's feedback for all the facilities provided in the public areas, UAE has introduced feedback kiosks for the visitors. The feedback request machine is a part of judging the Dubai happiness index.

As the UAE has the highest consumption of water per capita in the world, with the domestic consumption of only 13%, while 80% of water is utilized for different greening projects. Sweet water is obtained from the desalination of sea water at a huge cost.

The region is trying to use smart techniques to reduce the consumption of sweet water and reduce the cost of desalination and yet meet all the necessary water demands for the country. After meeting the demands of the population of UAE, landscaping became the second major priority for maintenance.

The UAE Government recognises that the nation's security is a collective responsibility and that individuals can potentially help in realising security and preventing crimes. Security services within the public open spaces or activity areas that enable people to communicate and report any suspicious or harmful incidents directly to the authorities. There are various types of smart devices and security provisions in the public spaces that always keep the users safe and secured.

The region has proved to be smart in the implementation of guidelines are prepared and made accessible through the municipality and authority websites and the incorporation of the mandatory regulations are periodically followed up by the Authority during the property licence renewal. The approval process of the new property development is complete after a thorough review of the considerations of the regulations.

As part of its vision 2020, the UAE government has earmarked billions to promote and develop urban greenery, through landscaping and horticultural projects, in a bid to transform the emirate into a "green paradise". There are conferences and activities arranged by the government that helps people of UAE and outside to share their experience and implement in smart practices in land scape architecture.

For a modern 'smart city' sustainable ecological landscaping is an important theme. "The use of 'smart' solutions, including ¬innovation within the landscape industry, and the increased use of technology, is making it easier for urban planners and environmental specialists to -implement sustainable, ecological landscaping initiatives. Environment Vision 2030 was developed in close coordination with the General Secretariat of the Executive Council and Department of Economic Development/Abu Dhabi Council for Economic Development to ensure integration among the environmental, economic and social visions, the three pillars of sustainability.

LEED has become popular in the industry and has even influenced the current attempts to develop a local framework.

- Conclusion

UAE will continue to expand and conserve the green open spaces though various innovative efforts and will set an example for the other countries. In moving towards the change, the building efficiency and sustainability play a vital part of the entire smart city concept—but there are also individual national, economic and cultural needs which must be met by technological developments.

Adaptability and malleability are other key factors in smart design and future projects need to strike a balance and be flexible between the built and open environment creation.

A smart city starts with mobility and creates a place where an individual feel safe. Data storage, data protection and data privacy are also a part of smart design.

Hence a holistic approach with a collaborative movement involving all stakeholders—including the research and development community, government authorities, developers and the general public is necessary for forming and running a smart city. Undoubtedly the benefits of a society that is seamlessly synchronized with technology will not only enhance but also evolve our living environments.

With the visionary leaders of UAE, progressive nature and love for the country, UAE is continuously making progress in every field ensuring the happiness factor working well within the region.

## References


1. T.M. Vinod Kumar, Associates (ed) (2014) Geographic information systems for smart cities. Copal Publishing Group, New Delhi
2. Vinod Kumar TM (ed) (2015) E-governance for smart cities. Springer, Singapore
3. Vinod Kumar TM (ed) (2016) Smart economy in smart cities" international collaborative research: Ottawa, St. Louis, Stuttgart, Bologna, Cape Town, Nairobi, Dakar, Lagos, New Delhi, Varanasi, Vijayawada, Kozhikode, Hong Kong. Springer, Singapore
4. Vinod Kumar TM (ed) (2017) E-democracy for smart cities. Springer, Singapore
5. Vinod Kumar TM (ed) (2018) Smart metropolitan regional development: economic and spatial design strategies. Springer-Nature, Singapore

The manufacturer's authorised representative in the EU is Springer Nature Customer Service Centre GmbH, Europaplatz 3, 69115 Heidelberg, Germany. If you have any concerns regarding our products, please contact ProductSafety@springernature.com

Printed and bound by CPI Group (UK) Ltd, Croydon, CR0 4YY
30/09/2026
02225556-0002